Design and Development of Kno
Management for Manufacturing

K. Ganesh • Sanjay Mohapatra
S. Nagarajan

Design and Development of Knowledge Management for Manufacturing

Framework, Solution and Strategy

Springer

K. Ganesh
SCM Center of Competence
McKinsey Knowledge Center India
Gurgaon, Haryana, India

Sanjay Mohapatra
Xavier Institute of Management
Bhubaneswar, India

S. Nagarajan
Department of Mechanical Engineering
Vickram College of Engineering
Enathi, Tamil Nadu, India

ISBN 978-3-319-35211-4 ISBN 978-3-319-02892-7 (eBook)
DOI 10.1007/978-3-319-02892-7
Springer Cham Heidelberg New York Dordrecht London

Printed on acid-free paper

Springer is part of Springer Science+Business Media (www.springer.com)

Preface

Knowledge is considered to be the learning that results from experience and is embedded within individuals. Sometimes the knowledge is gained through critical thinking, watching others and observing results of others. These observations then form a pattern which is converted in a 'generic form' to knowledge. This implies that knowledge can be formed only after data (which is generated through experience or observation) is grouped into information and then this information pattern is made generic wisdom. However, dissemination and acceptance of this knowledge becomes a key factor in knowledge management. The knowledge pyramid represents the usual concept of knowledge transformations, where data is transformed into information, and information is transformed into knowledge. Many organizations have struggled to manage knowledge and translate it into business benefits.

Implementing a KM system can be complex and dynamic, no matter how well planned and developed. Inevitably, a degree of organizational inertia is focused on the current rather than the new. Within an enterprise, people (personal and group) involvement and interests, process status and technology landscape can deflect the commitment needed to successfully implement such a system. Cumulative evidence from past research in KM suggests that effective implementation of KM solution in any organization requires robust designs and models for various critical elements of process, people and technology.

This book addresses modules/elements which are required before the implementation of KM solution in typical manufacturing and service industry. The objective is to develop generic framework, generic design and generic model for all the modules/elements and also to implement the same in a case study organization. The results from that have been provided as a solution to the KM problem described above.

Distinct Features of this book

- The proposed book has literature on how to design KM strategy which will align objectives of KM initiatives with business goals; this alignment is not shown in any other book.

- The book talks about a framework for KM implementation. The cases covered in the book have global appeal. No other book has proposed that type of framework.
- Usage of KM in manufacturing sector has been highlighted in this book, which no other book deals with.
- Case studies are provided that will show approach to design and implementation for KM strategy.
- Case studies have a global appeal as they have been prepared with the global audience in mind.

Chapter Contents of the Book

Chapter 1 deals with objectives, usefulness, drivers and processes of KM and importance and implementation of *KM*. The second chapter explains how KM has been evolved over a period of time and how it has been used. The chapter also details taxonomy and architecture used for KM design and implementation in different business houses. The third, fourth, fifth, sixth and seventh chapters discuss KM implementation issues and how to address them in a systematic manner. The chapters discuss theory, concepts, process architecture and models for implementing KM solutions. The eighth chapter explains role of balanced scorecard in designing key result areas (KRAs) for executives that will ensure accountability in implementing the KM solution. The last two chapters explain the role of vendor management in KM implementation and how to mitigate risks associated with implementation.

Gurgaon, India — Ganesh
Bhubaneswar, India — Sanjay Mohapatra
Enathi, Tamil Nadu, India — Nagarajan

Contents

Abbreviations

a	Regression coefficient
ABC	Activity based costing
APIOBPCS	Automatic pipeline, inventory order based production control system
B	Mean
BOM	Bill of materials
BSC	Balanced scorecard
BVQI	Bureau Veritas Quality International
CEO	Chief executive officer
CKO	Chief knowledge officer
CoP	Communities of practice
CPU	Central processing unit
CRM	Customer relationship management
CSO	Corporate strategic objective
DI	Delivery indent
DB	Data base
DFD	Data flow diagram
DLD	Deep learning design
DSS	Decision support software
e-commerce	Electronic commerce
ECM	Engineering change management
EDI	Electronic data interchange
EFQM	EFQM excellence model
ERP	Enterprise resources planning
ESICO	Electronics signals and controls
ETF	Electronic transfer of fund
EVM	Earned value management
FAQ	Frequently asked question
FEA	Functional economic analysis
FMEA	Failure mode and effect analysis
GPRA	Government Performance and Results Act
GDP	Gross domestic product

GUI	Graphical user interface
H	Hypothesis
ISO	International Organization for Standardization
IT	Information technology
IC	Investment company
IP	Internet protocol
IMRL	Innovative Manufacturing Readiness Levels
JSP	Java server pages
KBV	Knowledge based view
KC	Key characteristics
KM	Knowledge management
KMS	Knowledge management systems
KPIs	Key performance indicators
K-products	Knowledge products
L-CNO	Learning Collaborative Networked Organization
LO	Learning organization
MAKE	Most Admired Knowledge Enterprises
MCS	Management control systems
mKM	Mobile knowledge management
MRP	Material requirement planning
MRN	Material receipt notice
OL	Organizational learning
OS	Operating system
p	Probability value
P&G	Proctor and Gamble
PM	Performance measurement
PMI	Performance measure indicator
PMS	Performance measure system
POS	Point of sales
QAS	Quality at source
R&D	Research and development
R2	Coefficient of regression
RBV	Resource based view
RDBMS	Relative data base management
RFID	Radio frequency identification
RTC	Reaction to change
ROP	Reorder point
SCM	Supply chain management
SCOR	Supply chain operations reference
SME	Small and medium enterprise
SPSS	Statistical package for social sciences
SOA	Service oriented architecture
SAP	System analysis and program development
QAS	Quality at source
t	Significance variable

TEL	Technology enhanced learning
TQM	Total quality management
VMI	Vendor managed inventory
X	Independent variable
Y	Dependant variable

List of Figures

List of Tables

Chapter 1
Introduction

1.1 Knowledge Management (KM)

Public and private organizations all over the world have recognized the importance of knowledge to their long-term survival and success. Academicians and practitioners observe that knowledge is one of the most important sources of competitive advantage and the greatest asset for organizations. Boisot (1999) observes that we live today in a postindustrial society in which knowledge has increasingly come to be recognized as a primary source of wealth. He further states that the evidence has become overwhelming that economies that are poor in natural resources but skilled in the production and exploitation of knowledge generally outperform economies that have abundant natural resources but are lacking in such skills.

The 1990s saw a radical shift in the way organizations viewed themselves (and also their competitors). They realized that their competitive advantage lay more on their intangible resources like core competence, intellectual capital, ability to learn, and most importantly the collective knowledge of the organization. Thus, the basis of competition changed from traditional sources like land, labor, and capital to knowledge-based sources. Moreover, organizations realized that their products and services are nothing but the manifestation of their collective knowledge and also that knowledge is the principal driver of all other competencies and capabilities. Thus, organizations, world over are seeking ways and means to leverage their collective knowledge to create value for their customers and stakeholders. The past decade has seen the emergence of "practices" (both formal and informal) within organizations to "manage" this resource. Organizations (across several industry sectors) have evolved their own strategies for "knowledge management (KM)" that suit their vision, competence, and culture. Thus, the discipline of KM has grown in terms of both theory and practice.

The emergence of knowledge as an important factor of production can be attributed to the recent works in the area of strategic management and economic theory, especially, the resource-based view (RBV) of organizations. The RBV focuses on the resources and capabilities of organizations and not on its products and services

K. Ganesh et al., *Design and Development of Knowledge Management for Manufacturing: Framework, Solution and Strategy*, DOI 10.1007/978-3-319-02892-7_1,

(Barney 1991). Resources and capabilities can be thought of as a platform from which the firm derives various products for various markets. Leveraging resources and capabilities across many markets and products is the key in the resource-based strategy. While products and markets may come and go, resources and capabilities are more enduring. Thus, according to the RBV, competitive advantage based on resources and capabilities is potentially more sustainable than that based solely on products and market positioning (Zack 1999).

The RBV also advocates firms to possess "inimitable resources" that will enable them to sustain their competitive advantage. Academicians and managers agree that a company's knowledge may be the one thing that allows it to be competitive because all other resources are to a large extent reproducible. According to Zack (1999), companies having superior knowledge are able to coordinate and combine their traditional resources and capabilities in new and distinctive ways, providing more value for their customers than their competitors can. The literature on "organizational learning" (Argyris 1992; Senge 1990), "core competence" (Prahalad and Hamel 1990), and "intellectual capital" (Quinn 1992) have in many ways added to the importance of knowledge as a resource and underscored the need for organizations to build the capability to manage it.

The term "KM" was coined by Mr. Karl Wiig in 1986 at a conference in Switzerland. He stated that KM is a systematic, explicit, and deliberate building, renewal, and application of knowledge to maximize an enterprise's knowledge-related effectiveness – returns from its knowledge assets. But, later Nonaka (1991) mapped the term KM in the management literature. The famous quote "knowledge-creating companies" was emphasized and established by Nonaka (1991). The transformation of knowledge is explained in Fig. 1.1.

The major inspirations were drawn through the inspiration from the KM practices of firms like Matsushita and Canon. Nonaka and Takeuchi (1995) have produced a classic work in KM by expanding the theme of the "knowledge-creating companies." Leonard-Barton (1995) has triggered the KM revolution with a different theme called "wellsprings of knowledge." The KM practices of Chaparral Steel motivated the work of Leonard-Barton (1995). Post-1995, there has been factually a detonation in the literature on KM, including articles, books, and journals. In the year 1996, the *Strategic Management Journal* published a special issue on KM. *The Journal of* KM was launched in the year 1998.

Management consultancies like KPMG and Ernst & Young bestowed their part through several KM surveys (KPMG 1998) and distinguishing leaders in KM (MAKE 1998) among organizations. Many firms appointed chief knowledge officers (CKO) at the organizational level, similar to chief financial officers and chief information officers. The academia also witnesses the appointment of a "professor of knowledge" in the University of California. Gamble and Blackwell (2001) depicted KM in different dimension and stated that KM draws from a wide range of disciplines and technologies including cognitive science, expert systems, library and information science, organizational science, and network technology. Thus, KM came into sight as a discipline in itself.

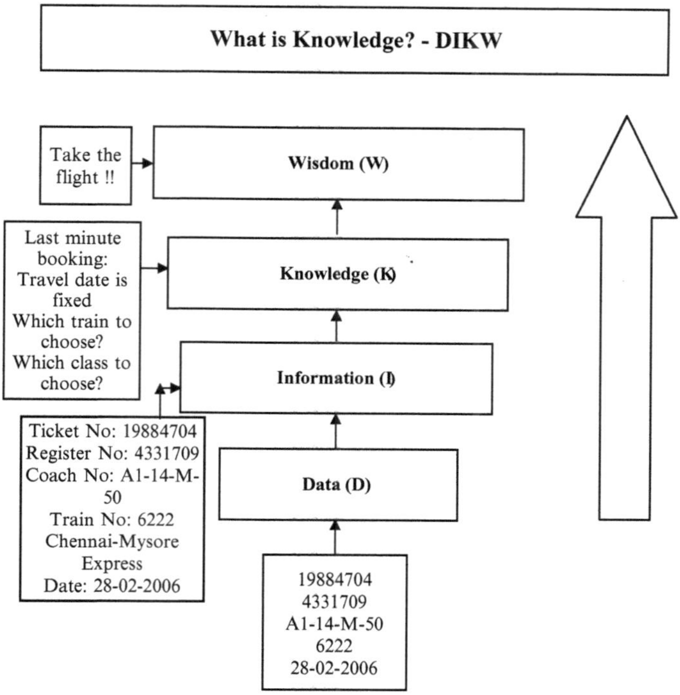

Fig. 1.1 Transformation of knowledge

1.2 Objectives, Usefulness, Drivers, and Processes of KM

The primary objective behind the KM initiative is to capture the explicit and tacit knowledge about people, skills, processes, markets, competitors, customers, suppliers, organization, environment, policies, procedures, regulation, legislation, etc. that exist in the organization in a structured manner and store the same as the organization's asset available to all employees on a "who need what basis." The forms of knowledge are clearly depicted in Fig. 1.2. Individuals, organizations, and nations have been managing knowledge for millennia. It is only in recent times that researchers and managers have realized the need for "conscious" and "purposeful" management of knowledge. Hence, the focus has shifted from managing knowledge embodied in various forms (like products, processes, patents, procedures, and in people) to manage knowledge itself as a resource and a capability. The reasons for this shift can be attributed to reasons within and outside organizations. Through a semistructured interview conducted with 43 executives belonging to 32 manufacturing organizations, the key drivers for KM were derived, which are detailed in Fig. 1.3.

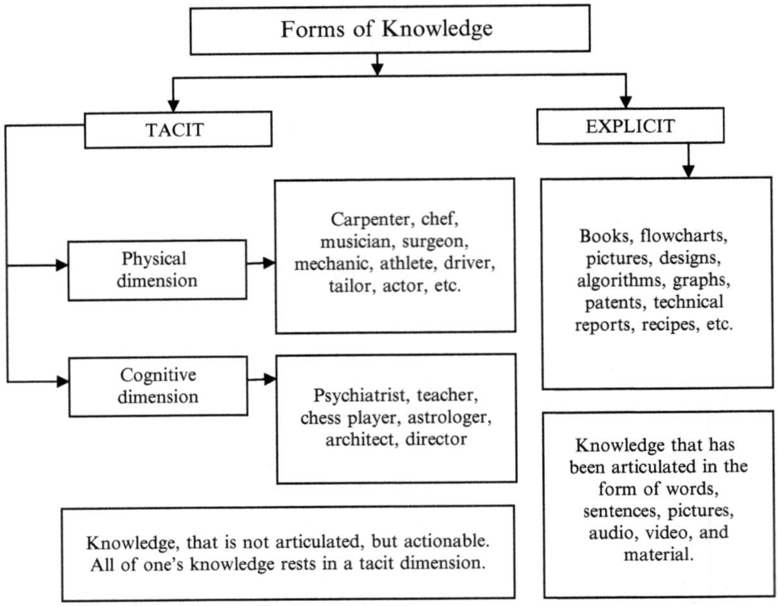

Fig. 1.2 Forms of knowledge

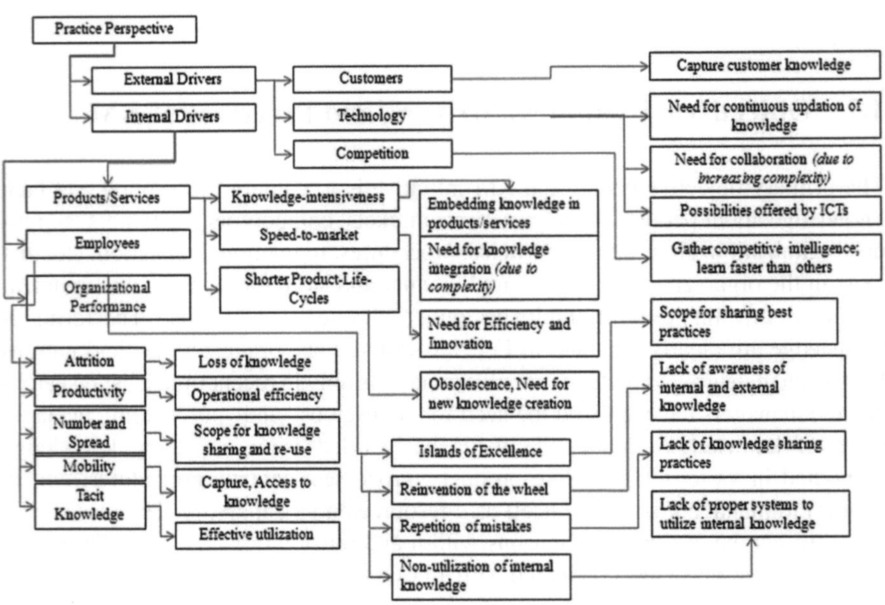

Fig. 1.3 Key drivers of KM

From the practice perspective, there are two categories of drivers, namely, internal and external drivers. The detailed subclassification of both internal and external drivers is collected from the executives based on their experience at the organization. The application of this knowledge in the workplace is to reuse knowledge to reduce rework, redeploy knowledge to leverage best practices, transfer skills and behaviors, repurpose knowledge to drive innovation, and achieve business benefits. KM also helps in tracking and retaining knowledge and information within the organization to provide it to the appropriate audiences in the most effective manner for which it is important to develop a knowledge-sharing culture and mechanisms to support it. The critical business benefits of KM are improved ability to capture and manage intellectual assets, effective dissemination of knowledge through collaboration, greater agility in responding to market and regulatory change, and improved knowledge continuity during organizational change.

KM involves managing the following knowledge-related activities in the organization:

1. Knowledge identification
2. Knowledge acquisition
3. Knowledge representation
4. Knowledge classification
5. Knowledge creation (or generation/production)
6. Knowledge capture
7. Knowledge storage
8. Knowledge utilization (or application)
9. Knowledge access (and retrieval)
10. Knowledge sharing
11. Knowledge transfer
12. Knowledge dissemination
13. Knowledge networking
14. Knowledge integration
15. Knowledge organization

The list of activities given above is neither mutually exclusive nor completely exhaustive nor is in a sequential order. Though the list contains 15 knowledge-related activities that an organization executes (either consciously or subconsciously), the main "KM Processes" emphasized both in theory and practice are:

1. Knowledge creation
2. Knowledge utilization
3. Knowledge sharing
4. Knowledge transfer

All other processes such as – identification, acquisition, representation, classification, capture, storage, access, dissemination, networking, integration, and organization – support the above four KM processes. It can be argued that if an organization consciously manages the above four KM processes, it would take care of all the other processes also. The key KM processes are detailed in Fig. 1.4.

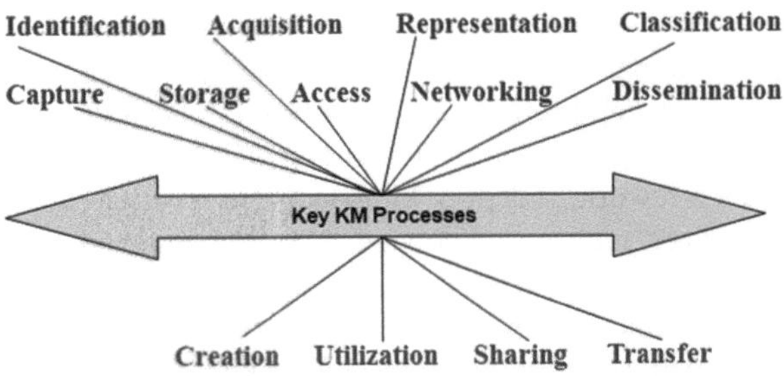

Fig. 1.4 Key KM processes

1.3 Importance and Implementation of KM

Cumulative evidence from past research in operations management and other disciplines suggests that effective communication and KM are critical elements of successful supply chain coordination and integration in manufacturing organization. "KM is defined as the organized and systematic process of generating, creating and disseminating information, and selecting, distilling, deploying and exploiting explicit and tacit knowledge through the critical pillars such as people, process, technology to create unique value that can be used to achieve a competitive advantage in the marketplace by an organization" (Nonaka and Takeuchi 1995; Pagell 2004). The importance of KM pillars is discussed as "KM addresses policies, strategies, and techniques achieved through people, process and technology aimed at supporting an organization's competitiveness by optimizing the conditions needed for efficiency improvement, innovation, and collaboration among employees" (Sousa and Hendriks 2006).

Unlike other management practices there are no standard guidelines or step-by-step procedures to implement KM in organizations. As yet, there are no "KM Maturity Models" or "KM Certification Agencies." This is because both the theory and practice of KM is still in the evolution stage. Many companies are experimenting with different models of KM. Since measurement of the costs and benefits of KM is also difficult, managers find it hard to convince their organizations to invest energies in KM. One of the advantages of KM is that it is wholly an "internal organizational affair," and external consultants cannot play a significant role in designing KM systems for organizations. Hence, people cannot argue that KM is just another management fad initiated by the academic, research, and consultant communities to make money for themselves. Given the vast amount of case studies, success stories, and frameworks available today on the Internet for free, all organizations need to do is form a KM team and set the ball rolling.

The following are some of the key points that need to be borne in mind when implementing KM in organizations:

People are the prime movers of KM. Unless all the employees believe in it, KM cannot succeed in the organization. It should neither be top-down driven or bottom-up driven. Employees should never feel that they are "driven" by someone; they should be self-driven. Thus, KM should be everyone's responsibility.

KM can be implemented through a big-bang approach or a piecemeal approach. In the big-bang approach, KM is an organization-wide process, whereas in the piecemeal approach, only a department or a function is taken for implementation. Many organizations have first experimented with a small division in their organization and then expanded it to other parts.

It is very important to integrate KM processes like acquisition, creation, sharing, utilization, and transfer into the core activities of the organization. Employees should never feel that they are doing "something extra" apart from their normal routine. Over a period of time KM activities will then become a part of the fabric of the organization. If KM is carried out as a "separate activity" removed from the daily routine, there is bound to be some resistance from employees. A dedicated KM team can be formed to take care of activities that are different from the organization's regular routine.

In any management activity measurement is an important part. Managers never believe in any concept or philosophy unless they can measure it with the instruments that they are familiar with. Before initiating KM, it is better to select an "organizational performance parameter" that needs to be improved using KM. This can range from reducing employee attrition, increasing sales in a particular region, improving a particular quality parameter of a product, reducing wastage, or increasing customer response time. The performance parameter can be measured before and after implementation of KM. This way KM will get integrated into the routine activities and also will gain confidence of the employees.

There is a common belief that KM is only the prerogative of big organizations with heavy investments in IT. This is not true. The size of the organization, number of employees, nature of business, infrastructure facilities, etc. don't matter for KM. Hansen et al. (1999) suggest two generic KM strategies – personalization and codification. Personalization strategy stresses on connecting people to people, i.e., people who possess knowledge to those who need it. Consultancies like McKinsey adopt this strategy. In codification the stress is on documenting knowledge, storing it in databases, and then disseminating it to people who need it. Consultancies like Ernst & Young adopt this strategy. Depending on the resources available and the culture, organizations can adopt either of these strategies for KM.

Some of the key KM activities that organizations undertake as part of KM implementation are given below. It should be noted that these activities are not presented in any chronological order. Though there may be interdependencies between certain activities, the KM theory has not yet devised a step-by-step guideline to implement KM:

1. Knowledge audit: The KM team consisting of key organizational employees who have sufficient knowledge of all the operations perform a "knowledge audit." This involves identification and cataloguing of key knowledge assets and knowledge competencies, identification of experts in various knowledge domains within the organization, and classification of knowledge assets into groups like "general," "strategic," and "proprietary."
2. Knowledge vision: The "knowledge roadmap" or "knowledge vision" document specifies the existing knowledge capabilities of the organization and the capabilities that need to be acquired in the short term and long term to remain competitive.

3. Knowledge mapping: As discussed earlier, this exercise involves "mapping" all the knowledge assets that exist in the organization. Knowledge maps can be constructed for specific departments/function/domains or for the whole organization.
4. Knowledge architecture: Just as the construction of a building starts with the plan or architecture, organizations need to design their architecture for KM. This is especially important for managing explicit knowledge predominantly in the form of documents and reports in databases and file cabinets.
5. Establishing knowledge roles and skills: This involves selecting and appointing a dedicated team of company personnel to oversee the knowledge activities of the organization. The roles and deliverables of the KM team should be specified clearly.
6. Knowledge infrastructure: Depending on the nature of KM issues and needs coupled with the resources available, organizations can invest in suitable infrastructure facilities including talk rooms, open spaces, canteens, libraries, computers, telephones, intranet, groupware, and Internet.
7. Organizational structure: Suitable modifications should be made to the existing organizational structure and systems that enables employees to implement KM activities without hassles. The structure should inspire a culture of KM.
8. Creating enabling contexts for KM: Many researchers stress that the important job of managers and KM activists is to create "enabling contexts" for KM. These "contexts" can be in the form of knowledge fairs, talk rooms, communities of practice, awards and incentives, idea contests, brainstorming sessions, and open houses. These contexts enable employees to open-out and freely express their ideas and exchange their experiences.

1.4 Motivation of Research

Implementing a KM system can be complex and dynamic, no matter how well planned and developed. Inevitably a degree of organizational inertia is focused on the current rather than the new. Within an enterprise, people (personal and group) involvement and interests, process status, and technology landscape can deflect the commitment needed to successfully implement such a system. To successfully implement a KM solution, six critical success factors must be considered and are discussed below (Bixler 2002):

1. Vision and leadership – KM strategic plan: A clearly defined and inspirational statement for the value of what the KM system is intended to accomplish within the enterprise. Success in implementing a KM solution within an enterprise relies on a well-designed KM strategy and an implementation approach tailored to the enterprise and its constituents.
2. Organizational and communication assessment and training: Organizational processes, procedures, and workflow are central to successful enterprise operations. The KM implementation strategy must be aligned with these organizational distinctions.

Additionally, a well-developed enterprise communication system and learning system is necessary to facilitate effective KM implementation. Emphasis must be on the critical importance of collaboration.

3. Business performance measurements: A KM system must absolutely be led by business drivers based on client demands, even if the enterprise itself does not recognize that. Leadership must define the business vision by demonstrating how a KM system will improve business processes and transform the enterprise. It needs to be recognized that KM implementation is not "business as usual."
4. KM mission interface and alignment: The scale of KM demand requires simultaneous, coordinated activities within an enterprise and a foundation of unification based on the organization's management structure. This ensures that it is possible to achieve the overall business vision of a KM system without compromising current service levels and existing business.
5. KM architectures and infrastructure: KM tools and the required infrastructure are essential and central to a successful KM system – but it is not everything. KM technology tools are the enablers of a KM system.
6. KM integration and resourcing: KM system integration within the enterprise is essential and requires proper resourcing. Senior management must commit to proper and practical levels of resource allocation to effectively manage and maintain a KM system and integrated program.

Cumulative evidence from past research in KM suggests that effective implementation of KM solution in any organization requires robust designs and models for various critical elements of process, people, and technology. With this background, a Delphi study was conducted (Nevo and Chan 2007a) with a semistructured questionnaire with top management executives those with decision-making authority, who select and evaluate the KM solution (KMS) of manufacturing organization in order to understand the set of critical elements which influence the KM solution implementation. Overall, 43 executives participated in the study, from various manufacturing industry in India varying in size from 200 to over 20,000 employees. Interviews were conducted in person and also over the telephone. All interviews were conducted based on the interview protocol. After each interview, notes were reviewed to identify potential challenges or problems. Only in one case, it was necessary to contact the respondent again for purposes of clarification. Four rounds were conducted to arrive to the final results.

Some of the statements from the interview are stated below:

- Well, I guess my perception is yes they have influence, but it's very labor intensive and very time intensive.
- It's doing what we asked it to do, but we're missing the boat on its potential right now, so that [is] actually something I say have influence.
- We're still too early to evaluate whether or not the million dollars worth of savings are real, but I honestly believe it is there.
- We're going to be able to have them access a program that will be able to help them out right at the bedside and be able to look up best practices so we're on that track and we're going down that road but we're not there yet.

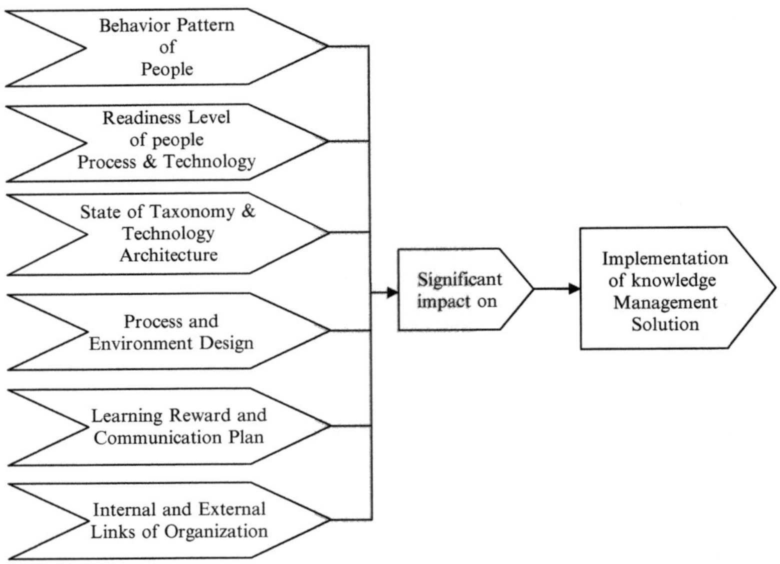

Fig. 1.5 Temporal confirmation conceptual model for research

- Where we are today and where we expect to be? I honestly think this will have influence.
- You believe that this is going to help you within your research to do things faster, then you have to be able to measure the time along the timeline – and that is a long, long time to measure, to be able to get through research, etc. so it's not a quick hit return that you're going to be able to see.

Based on the results of this Delphi study, a temporal confirmation conceptual model is developed (Fig. 1.5) (Nevo and Chan 2007b) with six modules/elements to take it to the next level of devising the generic framework, design, model, and strategy for the identified modules. From this study, it is clearly evident that the organizations should know the level and need a design and model for the below modules/elements before the implementation of KM solution:

1. Readiness level of people, process, and technology for the change
2. Behavior pattern of people for the change and for knowledge creation, use, and reuse
3. Taxonomy and technology architecture landscape with navigation and content layer of KM components
4. Process design for knowledge capture, storage and retrieval, and environment design for organization structure
5. Reward, learning, and communication design
6. Linkage design for internal and external levels and functions of organization

Keeping this as a base, a detailed business and research literature review is executed for all the six critical modules/elements to understand the research gap. Based on the detailed and thorough literature study and review on the six critical modules, namely, readiness assessment, behavior assessment, taxonomy and technology architecture, process, environment, reward and communication design, learning design and linkage design for internal and external levels, and functions of organization, the following points are observed:

1. The base research papers for readiness assessment are Hanafizadeh et al. (2009) and Khalfan et al. (2001). There is a need for conceptual framework and generic readiness assessment design for the readiness assessment exercise.
2. The base research paper for behavior assessment is King and Marks Jr. (2008). There is no generic behavior assessment model in KM perspective.
3. The base research paper for taxonomy architecture is Revilla et al. (2005). Generic base framework and model for KM taxonomy architecture is a research lacuna. Chua (2004) is the base research paper for technology architecture. Critical research gap is the generic technology architecture model for KM solution.
4. Paiva et al. (2008) is the base research paper for process, environment, reward, and communication design, and it reveals that there is no generic design for process, environment, reward, and communication typically for implementation of KM solution in manufacturing industry.
5. The base research paper for learning, reward, and communication design is Peter and John (2006), and a generic design and model for this element is a research gap.
6. Huang (2010) is a base research paper for linkage design for internal and external levels and functions of organization, and it indicates that the integration design for balanced scorecard and vendor-managed inventory are the critical research gaps.

1.5 Scope of the Present Study

1. Development of conceptual framework and generic readiness assessment design and model for the readiness assessment exercise in KM perspective
2. Development of conceptual framework and generic behavior assessment design and model for the behavior assessment exercise in KM perspective
3. Generic base framework, design, and model development for KM taxonomy and technology architecture
4. Development of generic framework, design, and model for process, environment, reward, and communication design
5. Development of generic learning design and model
6. Development of generic framework, design, and model for linkage of balanced scorecard and vendor-managed inventory in KM perspective

Chapter 2
Literature Review

2.1 Survey Overview

Rapid development of the Internet and information technology has pushed the world into the era of a new economy. In addition, with the revolution of information technology and the development of the Internet, the value of knowledge assets has been greatly enhanced. Armistead and Meakins (2002) argue that the creation of business value mainly comes from intangible assets, such as knowledge. The majority of studies show that knowledge can be classified as being either tacit or explicit (Nonaka 1991, 1994). Tacit knowledge is defined as experience-based knowledge that resides within an individual, whereas explicit knowledge is precise, formally articulated, and documented. In organizations, knowledge is often embedded in repositories, documents, routines, operational processes, practices, and norms. It is generally accepted that knowledge also comes from the meaningfully organized accumulation of information through experience, communication, or inference (Zack 1999).

Furthermore, knowledge activities are dynamic as well as humanistic with active and subjective natures created by social interactions dependent on individuals, their community and organization interactions, and applicability to needs (Holsapple and Joshi 2002). Hence, the activities of *KM* should enable the creation, communication, and application of knowledge, and they should drive the capability of creating and adding a greater value to the core business competencies. Several studies have proposed the concept of "knowledge gap" to describe the difference between the enterprise's current capability and the capabilities required for *KM* (Lovrich and Pierce 1984; Persaud 2001; Wild et al. 2002; Zack 1999). Nonaka (1991) stated that there exist different perceptions of *KM* activities and implementation among employees of differing levels and positions. The inability to identify and resolve any gaps prior to implementation greatly hinders the *KM* implementation. Thus, we propose a fully holistic framework of "KM gaps" to illustrate the management gaps that might occur when implementing the *KM* system. The reasons for these gaps and several fundamental approaches for avoiding them are presented. Through the evaluation of these gaps, enterprises can reduce the mismatch between the

K. Ganesh et al., *Design and Development of Knowledge Management for Manufacturing: Framework, Solution and Strategy*, DOI 10.1007/978-3-319-02892-7_2,

capability and implementation of the *KM* system and greatly enhance the effectiveness of implementation of the *KM* system.

For years, companies have strived to manage knowledge more effectively, the primary motivation being improved corporate performance (Choi and Lee 2002). However, despite the growing body of theory, there are relatively few *KM* texts that make an explicit connection between *KM* activities and corporate performance (Kalling 2003). Therefore, it is valuable to investigate how managers can eliminate *KM* gaps through *KM* activities in order to enhance corporate performance.

2.1.1 KM *in India*

India has a rich tradition of knowledge dating back to several millennia. It was our civilization that raised questions like "What is the ultimate reality?" and "How did this universe come about?" and so on. Indians had developed a fund of knowledge on various aspects including astronomy, metallurgy, food, health care, environment, plant and animal life, agriculture, linguistics, weapons, music, arts, governance, and ethics. This knowledge was disseminated both in the written form (in palm leaves) and through the oral tradition. The creativity of Indians can be evidenced from the fact that they "represented" the knowledge in various forms like prose, poetry, songs, stories (or puranas), proverbs, verses, slokas, sutras, metaphors, carvings, and pictures. In addition to these, much of this knowledge is embedded into the daily routine of people. Thus, the rich experience and wisdom of our ancestors has got integrated into our daily lives.

It is indeed ironical that a country with such a tradition has not taken the lead in *KM* in business. Both academicians and managers are responsible for this state of affairs. If we look at the *KM* literature (journal papers, magazine and newspaper articles, and books), there will be very few places where our country's name will be mentioned. This is despite the fact that we were the first to analyze concepts like *Buddhi* (intelligence), and *Vijnana/Jnana* (knowledge/wisdom). Our ancestors also separated *apara jnana* (knowledge concerning the created world) from *para jnana* (higher knowledge or knowledge of the ultimate reality). Our body of knowledge including the Vedas, the Epics, the Upanishads, the Puranas, the Brahma Sutras, and the Bhagavad Gita contain in-depth analysis of various concepts related to knowledge. While one may argue that our tradition did not address issues pertaining to business organizations, we need to remember the fact that the history of formal and organized businesses in India is not as old as those of the West. Thus, it is the responsibility of both academicians and practitioners to "adapt" concepts and experiences from other domains to business.

KM is still in its infancy in India. Very few companies have appointed dedicated personnel to take responsibility of *KM*. In most firms *KM* has been tagged on to somebody's existing responsibilities, often resulting in a step-motherly treatment. But this situation cannot last given the increasing competitive business environment in India. *KM* is no longer a luxury for Indian companies. It is a necessity that can make all the difference between survival and an early demise.

2.1.2 *Background for Detailed Survey*

Based on the study in Chap. 1, the organizations should know the level and process blueprint of the below elements before the implementation of *KM* solution:

1. Readiness level of people, process, and technology for the change
2. Behavior pattern of people for the change and for knowledge creation, use, and reuse
3. Taxonomy and technology architecture landscape with navigation and content layer of *KM* components
4. Process design for knowledge capture, storage and retrieval, and environment design for organization structure
5. Reward, learning, and communication design
6. Linkage design for internal and external levels and functions of organization

The detailed literature pertaining to all the modules/elements is discussed further in this chapter.

2.2 Readiness Assessment

Despite early proclamations regarding the value of knowledge and managing knowledge, the *KM* movement has been viewed as a "fad" or "recycled" concept (Spiegler 2000). The tide is turning, however, as more and more organizations become knowledge based. In the words of Davenport and Grover, "It is becoming increasingly clear that KM is here to stay. There are far too many knowledge workers dealing with too much knowledge for KM to disappear. Given that *KM* is becoming a competitive necessity, organizations and organization leaders find themselves asking 'Where to start?' and 'Is my organization ready?'" To begin, considerable qualitative research has suggested "enablers" of organizational *KM*. Although the literature is varied, common themes have emerged.

Implementing *KM* or knowledge-sharing projects in an organization requires significant organizational prerequisites. Lacking proper infrastructures and prerequisite not only make the *KM* process unprofitable but might incur harmful effects as well. To decrease such risks, it is proposed to introduce the readiness assessment, in order to gauge a company's appetite for the work involved in implementing the *KM*. The detailed review on readiness assessment in the context of *KM* is carried out and relevant literature is indicated here.

Kenett et al. (2008) applied multivariate methods in comparing risk profiles and readiness assessment at various stages of an enterprise system implementation. Li et al. (2010) presented the e-Health readiness assessment framework by integrating components of each reviewed framework and quantifying constructs within new framework. Ryoo et al. (2009) proposed a comprehensive readiness assessment framework to safeguards citizens and governmental organizations against the theft on their websites. Islam (2010) developed the readiness matrix called "innovative manufacturing readiness levels (IMRLs)" to understand and assess the maturity of micro- and nano-manufacturing technologies. Mohammadi et al. (2009) introduced the readiness assessment, in order to gauge a company's appetite for the work involved in implementing KM and used the exploratory factor analyses to validate the multi-item constructs. Holt et al. (2007) studied the *KM* and organizational change literature to take a first step in the development of a synergistic instrument that measures readiness for *KM* and applied it in an organizational setting. Shirazi et al. (2011) assessed employees' readiness to implement *KM* initiatives by examining the impact of individual, context, content, and process variables on commitment and pessimism prior to *KM* implementation. Keith et al. (2006) indicated statistically significant differences in *KM* readiness between groups and need for alignment by doing the field study from a Fortune 500 financial firm while transitioning its structure to service-oriented enterprises. Bui et al. (2003) illustrated and proposed a framework that can be used in providing e-readiness assessment and in making national strategic decisions on infrastructure which is conducive to the new economy. Miri-Nargesi et al. (2011) proposed a new model of readiness assessment factors of customer relationship management (CRM) and showed that top management commitment, project management capability, IT infrastructure management, customer-oriented culture, and clearly defined CRM processes are the top five readiness assessment factors.

2.3 Behavior Assessment

Literature holds that there exist two basic strategy approaches to culture in terms of implication: conforming strategy (maintaining order and continuity) and transforming strategy (changing and breaking existing patterns) (Bate 1994). Transformation in any organization happens as a result of an effectively managed change system. Consequently, a substantial segment of the change management literature focuses on describing how employees respond to change in organizations; how to handle the stress, conflicts, and emotional issues in the aspect of change in an organization; how to gain support to participate in the change effort; and generally, make organization-wide change less traumatic (Marquardt 1999; Edward 2000). Throughout this literature organizations are advised to recognize that change is implemented by and has consequences for people and that change can be made significantly less traumatic and more successful if these human aspects are anticipated and handled effectively. An entire literature has developed emphasizing the importance and impact of involving employees effectively in organizational decision making and change initiatives. Therefore it is crucial to assess the values, attitudes, and behaviors of the people in the organization that constitute barriers to seeking, sharing, and using knowledge before employing knowledge base and KM. Moreover, management of any organization needs to intentionally and carefully create conditions and stimulate the behavior needed for efficient knowledge sharing among employees, hierarchies, and across functions. The knowledge hierarchy is explained by Haeckel and Nolan (1993) in Fig. 2.1.

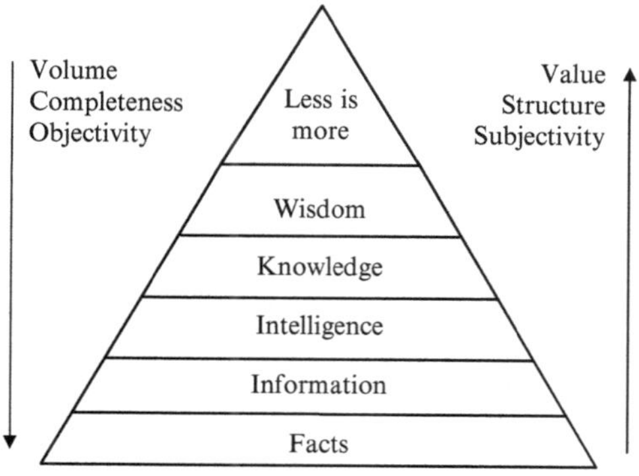

Fig. 2.1 From facts to wisdom

KM requires a collaborative approach from all management functions, across disciplines. Cumulative evidence from *KM* literature suggests that "KM addresses policies, strategies, and techniques achieved through people, process and technology aimed at supporting an organization's competitiveness by optimizing the conditions needed for efficiency improvement, innovation, and collaboration among employees" (Sousa and Hendriks 2006). They have also reported the impact of cultural and behavioral aspects in the domain of knowledge creation, shaping collectively crafted courses of action and addressing the methodology for the adoption of a knowledge-based view (KBV) in the organization.

Effective role of communication in KM have been found to be a critical element of successful supply chain coordination and integration in organizations (Nonaka and Takeuchi 1995; Pagell 2004). Nielsen (2006) proposed the dynamic capabilities approach to integrate KM activities and also identified the activities of knowledge creation, acquisition, capture, assembly, sharing, integration, leverage, and exploitation. Sousa and Hendriks (2006) have found that behavior of individuals is one of the critical components for *KM* solution implementation. Sabherwal and Fernandez (2003) proposed that socialization and combination influence perceived effectiveness of KM at group and organization levels, respectively. Germain and Iyer (2006) suggested that a task is already half complete if the integration in the organization is good among employees and top management. The top management has to enlighten members of the organization about the need for change, expressing the current status of the organization and where it needs to be in the future, and developing realistic approaches about how change might be accomplished. Thus available literature shows evidence of the role of people management in the organization for the effective setting up of *KM*. Further examination suggests that proper *KM* initiation can be gained through organizational culture that promotes organizational citizenship behavior. Podsakoff et al. (1990) examined the effects of transformational leader behaviors on organizational citizenship behaviors and the potential mediating roles of trust and satisfaction in that process. Organization leaders need to recognize that people in the organization are likely to resist making major changes for a variety of reasons, including fear of the unknown, inadequacy to deal with the change, and whether the change will result in an adverse effect on their jobs. It is also important

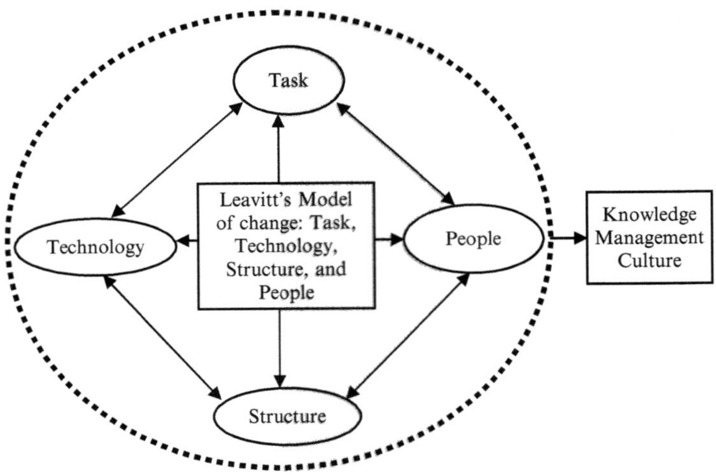

Fig. 2.2 Leavitt's model

to have trust and monitoring in an organization when any new change is initialized. Niehoff and Moorman (1996) have examined the theory concerning the relationship between monitoring and citizenship roles. It is also important to note that perceived fairness determines the full extent of cooperative contributions to organizations of citizenship behaviors (Organ and Moorman 1993). People need to feel that their concerns are being heard. Leaders must widely communicate the need for the change and how the change can be accomplished successfully. Choo et al. (2007a, b) suggested that quality of organization improves with knowledge-based system. Hwang et al. (2008) exhibited the method to gauge the actual performance and values of KM systems. It is also important for the employees to feel that the approach to change will include their strong input and ongoing involvement. How the employees perceive *KM* and the resultant change that occurs in an organization is also important. Nadler and Tushman (1988) have found that any change in an organization is effective only when it is fine-tuned, i.e., when it is anticipatory and incremental. Studies have shown that individual reaction to *KM* differs according to the way they perceive the change. Thus every system and mechanism in an organization is connected, and transformation is not that easy, especially if employees are not looped in an effective way. Change management can be effectual only when the interdependence of the mechanism is kept intact. The four interdependent elements prompting analysis and evaluation of interrelationships are shown in Leavitt's model of change which affects *KM* culture is shown in Fig. 2.2.

Hurley and Green (2005) have found the effect of task, structure, technology, and people characteristics on KM culture which is explained through a regression model and is shown in Fig. 2.3. Looking at the regression model of Hurley and Green (2005), it can be found that people characteristics have a regression coefficient value (0.40) indicating it as a strong predictor for KM culture.

Fig. 2.3 Regression model

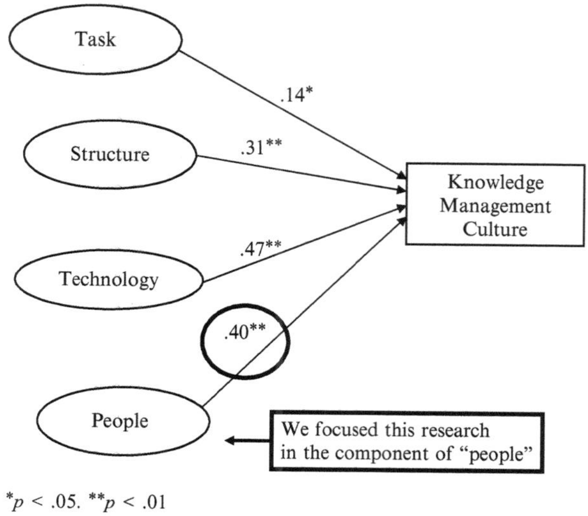

Researchers in organizational development (Lewin et al. 1939) have focused on the importance of a slow and steady adaptation and change management in an organization. The model of organizational development thrusts on the need for having a strong leadership and empowerment of its employees for a successful organizational transformation. The regression model used in this research also focuses on the need for concentrating on individuals rather than on technology, systems, and structure for a proper change management. Any organization can be regarded as a system, in which employees are the most important subsystem and that employees are the people who make structure and technology work, and therefore they become the cornerstone for the success of any organizational transformation.

A change could be taken as positive or negative depending on the perception of an individual employee. Any change would fundamentally impact a person's mental models of organizational life and subsequently would impact a person's attitude about the organization. Change affects everyone in a different way, but the actual challenge is to understand people's reaction to the alternation and to cope up with their responses. It is therefore imperative to understand and profile the change resilience of an individual employee. Effectiveness of behavior readiness in the organization toward the change management is shown in Table 2.1, which is adapted from the results of Ruben and Kealey (1979) and Chen and Huang (2007).

2.3.1 Relevant Research of Behavior in Aspect of Business Innovation and Research

Moutinho and Hutcheson (2007) developed a predictive model of store patronage behavior for consumers. Gao and Wei (2007) investigated the ethical acceptability

Table 2.1 Correlation model result data

	Respect	Interaction posture	Orientation to knowledge	Role behaviors: Empathy	Role behaviors: Task	Role behaviors: Relational	Role behaviors: Self-centered	Interaction management	Ambiguity tolerance
Culture shock	0.0000	0.446[2]	0.705[a]	0.543[b]	–0.153	0.687[c]	–0.483[2]	0.449[2]	0.417[2]
	(N=12)	(N=12)	(N=12)	(N=12)	(N=12)	(N=12)	(N=12)	(N=12)	(N=12)
Adjustment	0.686[a]	0.287	–0.004	0.209	–0.220	0.014	–0.222	0.377[2]	0.218
	(N=13)	(N=13)	(N=13)	(N=13)	(N=13)	(N=13)	(N=13)	(N=13)	(N=13)
Effectiveness	0.441[2]	0.483[b]	0.400[2]	0.349	–0.544[b]	0.059	–0.503[b]	0.308	0.443[2]
	(N=13)	(N=13)	(N=13)	(N=13)	(N=13)	(N=13)	(N=13)	(N=13)	(N=13)

of frequently used behaviors of Chinese enterprises from their executive's viewpoints. Zhao and Tian (2009) stated that consumer's perception of social rewards punishments influences their ethical judgement significantly in different consumption situations. Larsson et al. (2009) suggested a leadership chain model in which they identified the relationship between leadership behaviors, health, customer satisfaction, and profitable and effective organizations. Chen (2009) examined the joint impact of relationship-selling behaviors and switching barriers as a potential moderator in the light of current service quality, customer satisfaction, and customer loyalty paradigm. Singh (2009) investigated the relationship and the impact of organizational learning on to KM processes. Vouzas (2009) has drawn the attentions that the human resources issues such as human behavior and employee involvement and commitment are essential for the successful introduction and implementation for any processes. Suppiah and Sandhu (2011) indicated that the organizational culture types influence tacit knowledge-sharing behavior and that influence depending on the cultural type. Mills and Smith (2011) evaluated the impact of specific KM enablers and processes on organizational performance.

2.4 Taxonomy and Technology Architecture

According to Cote (2005), "Taxonomy – the classification of items within subject domains – is especially effective in helping with today's information access difficulties. These structures are particularly good at representing open systems and are useful in the visual world of the Web. They are helpful in portraying abstract concepts and reflecting the various forms knowledge can take, whether tangible or intangible. Taxonomies do not necessarily use pre-existing classification schemes and are often based on a synthesis derived from user need and language. Taxonomy can be hierarchical systems that also use controlled vocabularies and thesauri, but they are not always based on the accepted standards used in a traditional setting. The multifaceted subject headings of a taxonomy have the added bonus of reflecting connections and processes in a semantic structure, or ontology, allowing for potentially wider consultation and application of knowledge objects. Taxonomies leave room for growth and constant revision, depending on external or internal factors." The detailed review on taxonomy and technology architecture in the context of *KM* is carried out, and relevant literature is indicated here.

Doherty et al. (2005) described the *KM* techniques along with ontology to improve the design, implementation, and operation of the environment through expertise knowledge base system. Gupta et al. (2009) suggested that organizations have to infuse desirable values for knowledge sharing and have to align their practices conducive for knowledge sharing in organizations. Spiteri et al. (2010) contributed a mobile KM (m*KM*) system architecture aimed at providing designers situated in mobile work settings with life cycle on sequences knowledge support. Lai et al. (2009) investigated the structure of industrial network strategies in the current global manufacturing environment based on the empirical analysis and employed hierarchical analysis to devise a taxonomy for industrial network strategies. Reddi et al. (2011) proposed a service-oriented architecture (SOA)-based framework to carryout engineering change management (ECM) across a supply chain and to achieve effective ECM over a collaborative network of product development. Lynch et al. (2009) applied the taxonomic process to operational decisions using the Decision Support Software (DSS) for a key decision-making process in the organization. Lodree et al. (2009) established a framework for scheduling human tasks that account for physical and cognitive human characteristics and behaviors. Liu et al. (2008) described a *KM* and retrieval system based on hierarchical text classification scheme and organizes the large volume of manufacturing-related electronic documents according to manufacturing knowledge taxonomy. Snyder et al. (2000) stressed to identify the taxonomy of processes to communicate the *KM* process and framework of *KM* in the organization. Carthy et al. (2000) presented an evolutionary management technique "cladistics" for the organizations to formally and systematically understand the emergence of new manufacturing forms within their business environment.

2.5 Process, Environment, Reward, and Communication Design

The process design, organizational structure design, reward plan, and communication plan are the key elements before the implementation of *KM* solution. The detailed review on process, environment, reward, and communication design in the context of *KM* is carried out, and relevant literature is indicated here.

Choi et al. (2004) explored the framework for the *KM* and business process management and integrated the two paradigms into a single framework and also classified the process knowledge into three types. Rehman et al. (2007) explained the true meaning of design context knowledge that can be used in a structured way to support decision making and the prediction of their consequences at the conceptual design stage. Wei et al. (2008) examined the impact of *KM* process on innovation of firms and suggested that the knowledge process architecture of knowledge acquisition, integration, exploitation, and protection is essential for effective organization innovation. Hassan et al. (2007) explained the concept of reusing the knowledge about causalities and relations between Key Characteristics (KC) and validation of design robustness using Failure Mode and Effect Analysis (FMEA) knowledge. Franken et al. (2006) suggested that the choice of *KM* approach must be closely aligned with the organization's strategic and operational decision in order to get the anticipated benefits. Salah et al. (2011) presented a knowledge-based system to enhance creative conceptual design, and it requires the integration of many components, namely, design process, creative tools, and design knowledge. Gunther et al. (2008) presented two techniques for mining change logs in adaptive process management systems and showed the use of process mining as an analysis tool for truly flexible processes by understanding when and why process changes become necessary. Kunz et al. (2010) proposed the management control systems (MCS) have considerable impact on an individual's knowledge processes and the success of implementing KM in firms. Khalfan et al. (2010) highlighted the benefits of integrated construction SCM through effective *KM* and concluded that it will improve the overall production. Moon et al. (2011) suggested that communities of practices (CoPs) have been recognized as an effective vehicle for taking advantage of explicit and tacit knowledge within an organization.

2.6 Learning Design

Our intelligence is based on our learning community. Likewise organizational intelligence is the result of a collective intelligence and knowledge sharing. Levitt and March (1988) have held that organizational learning is a form of intelligence. Knowledge is organized in a way that its disciplines are integrative and interactive, which would ultimately lead to firm learning. Pedler et al. (1989) coined the term learning organization (LO) initially. Prietula and Simon (1989) developed an expert intuition through the pattern recognition over many years of experience. Senge (1990) popularized the term LO with his book *The Fifth Discipline: The Art and Practice of the Learning Organization*. Seely-Brown and Duguid (1991) pointed out that learning and improvisation occurs within communities of practice, and the communities can be either formal or informal, which becomes a backdrop of *KM* initiatives. Dyck et al. (2005) suggested that the tacit error correction phase in *KM* happens in the socialization process, largely as improvisations by the communities of practices.

Behling and Eckel (1991) suggested two types of intuition, namely, an expert intuition and entrepreneurial intuition, which would be imperative in knowledge sharing and successful *KM* implementation. Isaacs (1993) suggested that the cognitive map of the individual is both influenced by, and influences, the domain where the process takes place. Argyris (2004) proposed that a sociopsychological process of learning can result in either single-loop or double-loop learning, learning that occurs in organizational learning stage. Tsang (1997) examined the dichotomy between the prescriptive and descriptive researches and found that organizational learning happens only when the two streams of research are integrated. Crossan et al. (1999) proposed that organizational learning is the principal means of achieving the strategic renewal of an enterprise. Bontis et al. (2002) analyzed the social and behavioral dynamics of learning processes. Nonaka (1994) proposed a paradigm for managing the dynamic aspects of organizational knowledge creating processes, and Robinson (2001) distinguished the normative nature of LO from descriptive nature of OL, where in learning organization models emphasize leadership and management, culture, and systems for communication, information, and knowledge along with organizational learning.

Raman and Murali (2000) suggested that collaborative knowledge creation and sharing in an environment is inevitable for the success of the firm. Rahim (2002) pointed out that only when an interactive process manifested in incompatibility, disagreement, or dissonance within or between social entities will true learning happen. James and Abraham (2002) addressed the characteristics of learning in knowledge-based work environments of a software firm is the proper use of cooperative decision making. *KM* has a multidimensional nature at the organizational level while managing knowledge (Argote et al. 2003) and can be analyzed as the social and behavioral dynamics of learning processes (Bontis et al. 2002). Ambjorn et al. (2008) implied that the role of conceptual modeling is the key process for implementing a better learning design.

De Kereki et al. (2004) proposed a new model with learning environment which improved problem-solving skills and better capacity to transfer knowledge from one situation to another. Cavaleri (2004) explained the knowledge transfer has widely changed at the organizational level from 1st generation to 2nd and 3rd generation *KM* and that the process of *KM* integration happens successfully when knowledge transfer happens from individual and group level to the organizational level (Vera and Crossan 2004).

Hazlett et al. (2005) elaborated that the multidimensional nature of *KM* has evolved from computational view to organic view. Argote (2005) further explained that the organic view concentrates on tacit and explicit knowledge and more on the computational view. Masataka et al. (2006) explained that with practice and learning organizational capability for creating intellect can be increased and later developed an ontology-based *KM* environment. Kenneth (2010) revealed how the learning and KM principles are applied in internal processes and used modified learning model. It was Peter and John (2006) who linked the OL, LO, and *KM* at the process level and theorized the five disciplines of Singe and impacted the knowledge creation spiral of Nonaka, providing better insight to the concept and its interrelations. Sun and Scott (2003) found that some barriers are involved in transfer of learning to all levels in the organization and the absence of a link to the learning processes are identified as the major issues in implementation failures. It is postulated that these are the reasons for the gap between the two streams. Hans (2009) elaborated the obstacles of various nature which restrain to KM implementation and also the policy and decision making in view of cost–benefit. Andrew et al. (2010) elaborated an approach to designing technology-enhanced learning (TEL) and outlined the framework of Deep Learning Design (DLD) which leads to sustainable innovations in learning.

Pablos (2004) tried to analyze, from a KM view, how multinational corporations face the triple objective of achieving global efficiency, local sensibility, and organizational learning and found that human resource policies act as backbones to their success. Alas and Vadi (2006) conducted an empirical research in 44 Estonian organizations which indicated an impact of institutional environments on how cultural orientations influence organizational learning and result in inducing better employee attitudes toward change. Leseure (2010), through multiple case studies of Moroccan textile apparel offshore factories, stated that one organizational design and culture might not fit well within international operations networks. Alas (2008) compared the organizational change and learning in production and service sectors during a transition in society. From his interviews with members of top management teams of 137 Estonian organizations, it was understood that culture mediates the reaction of employees regarding change. Alas et al. (2008) conducted a survey in 29 Chinese organizations and also contributed to the role of OL and culture as inevitable in readiness to change. It reveals the importance of the task and relationship orientation of organizational culture to bring out organizational learning and to have better change readiness in the employees. Singh (2009) investigated the relationship and the impact of organizational learning on KM processes through a quantitative research investigation in which the participants were administered psychometric instruments on organizational learning and KM and found the relationship to be substantial. Loss et al. (2010) introduced a theoretical framework for Learning Collaborative Networked Organizations (L-CNOs). It merges both organizational learning (OL) and KM paradigms to support the main elements presented as part of the framework, which supports this present study.

2.7 Balanced Scorecard

The literature review for balanced scorecard (BSC) in the context of *KM* is detailed in Table 2.2 and Table 2.3

Table 2.2 Strategy tool for performance management system

Business metric	Contribution of research	Reference
Intangible valuation	Intangible assets are evaluated to extract their value components and align them with business strategy	Green and Ryan (2005)
Balanced score for BSC	Theory is used to calculate the relative weightage for each factor and comparison	Punniyamoorthy and Murali (2008)
New challenge for BSC	Organizations have quite different needs, market areas, people, products, and services and will end up with significantly different balanced scorecards	Chavan (2009)
Strategic planning	General concepts of strategies planning, system of management of quality and BSC, in order to assist it in the development of the model	Goncalves (2009)
Performance measures	To analyze the efficiency and benefits of supply chain (SC) scientifically and validate the usability of methods on performance evaluation index system	Yang (2009)
Mapping of *KM* and business strategy BSC with greenness	Impact and importance of *KM* activities in performance of BSC, such as financial, internal process as well as learning and growth perspectives	Chen and Mohamed (2008)

Table 2.3 Application for balanced scorecard

Application	Contribution of research	Reference
Finnish municipalities	Analyze of interrelations of the decision-making rationales around accounting performance measurement (PM) adoptions in Finnish municipalities	Rautiainen (2009)
Stock exchange	Analysis of BSC usage among companies on the Thai stock exchange	Kittiya and Guthrie (2009)
Electricity corporation	Importance of the rational analytical deliberation of legitimacy as a fundamental accompaniment to isomorphism in the continuing development of the new performance management system	James (2009)
Logistic industry	From a supply chain perspective, the non-tangible measures such as customer satisfaction are most measured	Chia et al. (2009)
Performance of SMEs	Integrated performance measurement framework for supply chain evaluation and planning in small and medium enterprises (SMEs)	Thakkar et al. (2009)
Performance of university	To maintain university operating standards, encourage individual universities to work on inadequacies, and promote university competitiveness, performance measure indicators (PMIs) was established	Chen et al. (2009)
Investment companies	A methodology for selecting strategic processes among the processes of Investment Company (IC) based on the BSC framework and the statistical analysis	Hanafizadeh et al. (2009)
Finish food manufacturing	The integration of performance measure system(PMS) and EMS issues into BSC	Lansiluoto and Jarvenpaa (2008)
Cooperative bank	Case study shows how the selection of performance indices affects results and the evaluation of a firm's performance	Chen et al. (2008)
Petroleum industry	Analytical hierarchy process (AHP) and balanced scorecard (BSC) for evaluating performance of the petroleum supply chain	Varma et al. (2008)
Performance in financial sector	To assess the improvements that relate to learning and growth, internal processes, and customers in financial performance	Cohen et al. (2008)

2.8 Vendor-Managed Inventory

In its simplest form, vendor-managed inventory (VMI) is the process where the vendor assumes the task of generating purchase orders to replenish a customer's inventory. VMI is a term that is used to describe many types of supply chain initiatives. These different "VMI" activities can vary substantially in purpose and application. In all of its forms VMI should be about improving visibility of demand and product flow in a supply chain, facilitating a more timely and accurate replenishment process between a supplier (vendor) and manufacturer or between a manufacturer and supplier. The VMI process is a combination of e-commerce, software, and people. The e-commerce layer is the mechanism through which companies communicate the data. VMI is not tied to a specific communications protocol. VMI data can be communicated via EDI or any other reliable communications method. The key feature of the e-commerce layer is that the data will be timely and accurate. The literature on VMI can be broadly classified as follows:

1. Mathematical/simulation models:
   (a) Single vendor – single retailer
      (i) Deterministic demand
      (ii) Stochastic demand
   (b) Single vendor – multiple retailer
      (i) Deterministic demand
      (ii) Stochastic demand
2. Empirical analyses

2.8.1 Single Vendor and Single Buyer with Deterministic Demand

Hoque and Goyal (2000) developed an optimal policy for single-vendor, single-buyer-integrated production inventory system with capacity constraints and equal and unequal sized shipments. Chaouch (2001) presented a VMI model seeking the best trade-off among inventory investment, delivery rates, and shortage. Fry et al. (2001) compared the Retailer Managed Inventory and VMI system and showed the latter is superior. Dong and Xu (2002) modeled the profit function of vendor and buyer. Viswanathan and Wang (2003) proved that the VMI model achieves perfect coordination with simultaneous offer of quantity and volume discount.

2.8.2 Single Vendor and Single Buyer with Stochastic Demand

Disney and Towill (2002) highlighted the evaluation and optimization procedure for Automatic Pipeline, Inventory Order Based Production Control System (APIOBPCS)-VMI model in order to minimize inventory holding cost and production adaptation costs. David and Eben-Chaime (2003) formulated and analyzed inventory cost models for each party and also for the joint systems. Disney and Towill (2003a, b) built a simulation model to analyze the bullwhip effect. Disney et al. (2004) established e-business-enabled supply chain models for quantifying the impact of information and communication technologies, particularly on the effect of dynamic behavior.

2.8.3 Single Vendor and Multiple Buyer with Deterministic Demand

Lu (1995) presented an optimum solution procedure for one-vendor one-buyer case and developed a heuristic approach for the multi-buyer case. Waller et al. (1999) modeled a manufacturing plant that supports multiple retailers through distribution centers of three different types and simulated the model for performance evaluation. Cetinkaya and Lee (2000) presented an analytical model for coordinating inventory and transportation decisions in VMI system. Achabal et al. (2000) modeled a decision support system for VMI with forecasting model, inventory decision model, and parameter estimation and updating model for an apparel manufacturer. Axsater (2001) considered the integrated inventory and transport decisions in VMI systems modeled by Cetinkaya and Lee (2000) and presented an efficient algorithm for exact optimization and an alternative heuristic. Viswanathan and Piplani (2001) proposed a model to study and analyze the benefit of coordinating supply chain inventories through the use of common replenishment periods. Piplani and Viswanathan (2003) conducted a numerical study to analyze how various parameters affect the total costs under supplier-owned inventory strategy. Smaros et al. (2003) used simulation to analyze the demand visibility.

2.8.4 Single Vendor and Multiple Buyer with Stochastic Demand

Kleywegt et al. (2002) formulated inventory routing problem as a Markov decision process and proposed approximation methods to find good solutions. Gerchak and Khmelnitsky (2003) modeled retailer's and supplier's problem in a newsvendor environment and used dynamic programming approach. Jaruphongsa et al. (2003) formulated a mathematical model and provided a polynomial time algorithm based on dynamic programming approach to solve the problem optimally.

2.8.5 Empirical Analyses/Case Studies

Holmstrom (1998) conducted a case study to show how VMI could be implemented for an enterprise resources planning (ERP) environment. Smaros and Holmstrom (2000) studied the data capture, data transfer, and data management in VMI system. Daugherty et al. (2001) conducted an empirical study, which focused on electronic data interchange (EDI), bar coding, and information system capabilities. Kaipia et al. (2002) developed a time-based method for measuring the benefits of VMI. Kulp (2002) studied the relationship between VMI, information precision, information reliability, consumer demand variability, and VMI use on wholesale price. Tyan and Wee (2003) examined the implications of VMI system in the Taiwanese grocery industry by means of a survey and concluded that VMI has the ability to reduce

Table 2.4 Applications of vendor-managed inventory

Application	Contribution of research	Reference
VMI in construction	Challenge of managing logistics in small items in construction industry	Tanskanen et al. (2008)
Patterns of VMI	Three empirically grounded patterns of VMI are proposed. Five contextual inhibitors of VMI impacts are suggested	Kauremaa et al. (2009)
Information exchange in VMI	Discrete event simulation is used to examine how a manufacturer can combine traditional order data available from non-VMI customers with sales data available from VMI customers in its production and inventory control and manufacturer efficiency	Smaros et al. (2003)
Retail industry	No of factors to make the efficient operation of VMI difficult, analyze the strategy is effective when the relationship between major retailers and major suppliers is constructive and open	Blatherwick (1998)
Guideline of VMI	Proposed standard agreement has a flexible structure easily adopted in industrial fields of VMI	Zammori et al. (2009)
Success factors of VMI	VMI success is impacted by quality of the buyer–supplier relationship, IT system, and intensity of information sharing	Marloes et al. (2008)
Dynamic inventory in VMI	VMI-based simulation approach can minimize the inventory level	Reddy and Vrat (2007)
Benefits of VMI in stochastic demand	Proposed VMI strategies, to reduce the order picking cost and transportation costs resulting in reduced total supply chain costs	Kiesmüller and Broekmeulen (2010)

costs, improve the service levels, and create business opportunities for both parties in the supply chain. Yang et al. (2003) identified five critical factors of VMI from literature and examined through detailed simulation of a hypothetical dual-level VMI delivery system. De Toni and Zamolo (2004) implemented VMI for electrical household appliances with focus on forecast and dispatch plan. Kuk (2004) tested the impact of VMI on information quality enhancement, service improvement and cost reduction through survey in electronic industries. The applications for vendor-managed inventory are detailed in Table 2.4.

2.9 Observations and Research Gap

A careful analysis of literature on the modules/elements which is critical for evaluation and assessment before implementation of KM solution hitherto reveals the following: it is evident that there is a need to develop generic framework, design, and models for all the six modules/elements. Given the strategic importance of knowledge, there is no doubt that organizations should make sufficient efforts to manage this critical resource. This is important both for their long-term survival and success and building sustainable competitive advantage. From the state of the art of literature, it is evident that the assessment and development of the below modules/ elements are not considered in the research:

1. There is a need for development of conceptual framework and generic readiness assessment design and model for the readiness assessment exercise in *KM* perspective.
2. There is a need for development of conceptual framework and generic readiness assessment design and model for the behavior assessment exercise in *KM* perspective.
3. Generic base framework, design, and model for *KM* taxonomy and technology architecture is a research lacuna.
4. There is a need for development of generic framework, design, and model for process, environment, reward, and communication design for implementation of *KM* solution in manufacturing industry.
5. A generic learning design and model is a research gap.
6. Development of generic framework, design, and model for linkage of balanced scorecard and vendor-managed inventory in *KM* perspective is needed.

The objective of this research is to develop generic framework, design, and models for all the modules/elements and to implement for a case organization to demonstrate the effectiveness and efficiency of framework, design, and models.

2.10 Summary

This chapter presents a survey of literature on the selected modules/elements related to *KM* which needs to be evaluated and assessed before the implementation of *KM* solution. The gaps identified from the literature provided the motivation for the chapters addressed in this book.

Chapter 3
Readiness Assessment for Knowledge Management Solution Implementation

3.1 Introduction

In order for *KM* to deliver value to the organization, it is essential for the organization to develop a number of pillars/dimensions. The *KM* strategy should ideally convert itself into actions in all areas. The primary target for *KM* will be creation, dissemination, and exploitation. However, it has to be very strongly supported by all the enablers to realize the desired value. This is clearly depicted in Fig. 3.1.

The strategy stage for *KM* will begin with a review of the organization's goals and objectives for *KM*. Once the objectives are understood, it leads to the development of knowledge components, which describes what the organization needs to know to accomplish its goals. Next to the opportunity and gap analysis is the state of readiness called knowledge maturity for any organization that can be achieved by systematically addressing and reviewing the three critical pillars of *KM* – people, process, and technology (Krogh et al 2001; Nonaka et al 2002; Siemieniuch and Sinclair 2004). The aim is to reach a state where it gets entrenched in the business processes by incessantly promoting the *KM* readiness. It is thus apparent that it is a path of continuous improvement and must be administrated by a strong readiness review approach, which has the ability to assess and benchmark the various aspects of people, process, and technology in a holistic manner. Readers can refer to Chen and Huang (2007) and Mrayyan et al. (2008) to understand the organizational readiness. This review determines how well the organization is positioned to adopt *KM*. In order to develop a sense of direction, the readiness assessment outcome should give an indication on how organization needs to adapt when the *KM* initiative is rolled out. This research focuses on the development of readiness assessment framework and approach. The study has taken the case of the Indian textile machinery manufacturing industry, and the readiness assessment framework and approach was developed for a typical organization and it is also generalized for engineering manufacturing industry.

K. Ganesh et al., *Design and Development of Knowledge Management for Manufacturing: Framework, Solution and Strategy*, DOI 10.1007/978-3-319-02892-7_3,

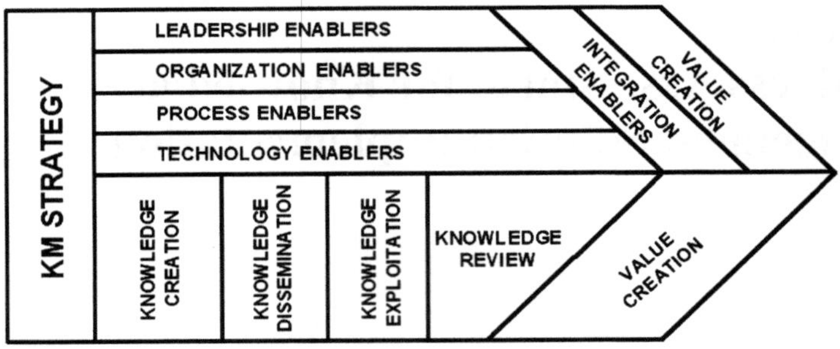

Fig. 3.1 Value chain of *KM*

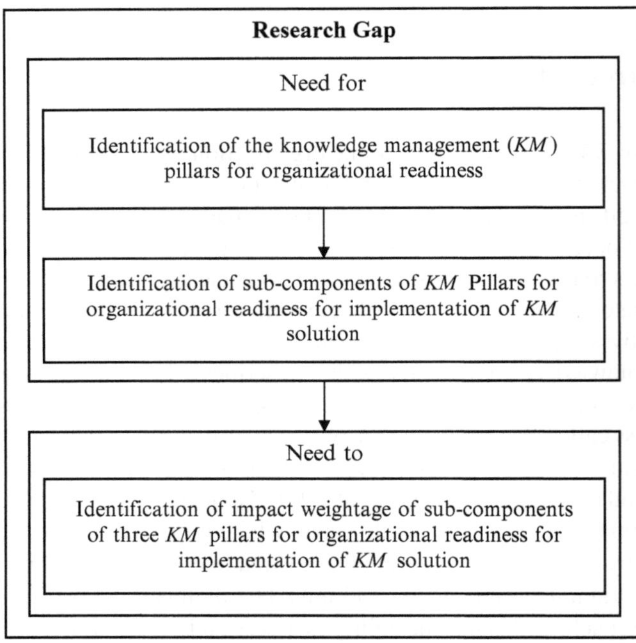

Fig. 3.2 Research gap

3.2 Research Gap Based on Literature

Cumulative evidence from past research in *KM* suggests that effective implementation of *KM* solution in any organization requires a readiness assessment of process, people, and technology and its elements. The primary intention of readiness assessment is to devise and design the readiness assessment framework, approach, and model for the implementation of *KM* solution. From the detailed literature survey, the research gap is shown in Fig. 3.2.

3.3 Research Process and Methodology

The readiness assessment framework, approach, and model for the implementation of *KM* solution are developed through this study.

The research process is divided into two phases:

Phase 1. Identification of *KM* pillars for readiness and development of readiness framework with subcomponents of *KM* pillars toward the implementation of *KM* solution

The phase 1 includes two steps:

Step 1 of Phase 1 – Identification of *KM* pillars for readiness based on business and research literature

Step 2 of Phase 1 – Development of readiness framework with subcomponents of *KM* pillars for organizational readiness through semistructured interview and Delphi study

Phase 2. Subcomponent impact for readiness

The phase 2 includes one step:

Step 1 of Phase 2 – Identification of impact weightage of subcomponents of three *KM* pillars for organizational readiness of *KM* solution implementation through questionnaire-based survey

This process is designed in such a way that any organization can use this as a base and they can develop organizational readiness framework based on the three pillars of *KM* and identify the impact of main and subcomponents of *KM* pillars for organizational readiness in order to improve the readiness level of *KM* pillars in the organization for the implementation of *KM* solution.

The research methodology used in the research process is detailed here:

In Step 1 of Phase 1, exploration of business and research literature is the methodology to identify the *KM* pillars. Organizations seek to develop a competitive advantage in market through reduction of lead time, reduction in cost, and improved productivity. However, the market environment is dynamic and the issues of globalization, rapid technology diffusion, and dearth of quality human capital resources require an exemplar shift in the approach toward strategy management and development. In an economy where the only certainty is uncertainty, the one certain resource of lasting competitive advantage is knowledge. *KM* seeks to improve an organization's usefulness by leveraging the knowledge it has, to improve its core proficiency. When markets shift, technology flourishes, competitors proliferates, and products become superseded, successful organization are those that constantly create new knowledge, propagate it widely throughout the organization, and quickly exemplify it in new technologies and products. Success in such a highly dynamic environment requires that organizations are more receptive to their customers, more agile in the way they do business, and more focused on core competencies through the support of people, process, and technology. From the business and research literature, it is evident that the three pillars along

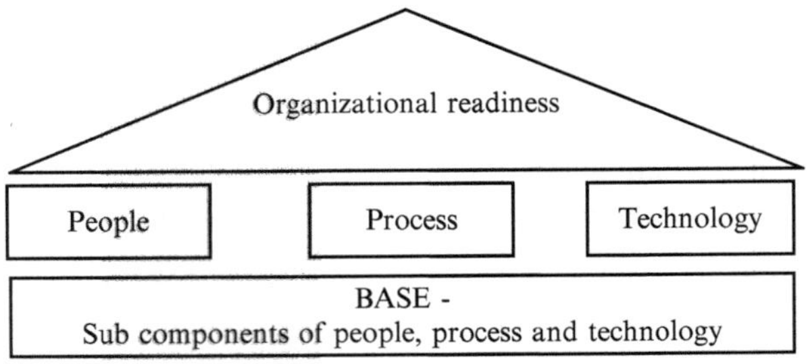

Fig. 3.3 Pillars of *KM* for organizational readiness

with its subcomponents are solely responsible for the readiness of organization toward the implementation of *KM* (Fig. 3.3). These three *KM* pillars can be used as a base for any manufacturing organization.

In Step 2 of Phase 1, a Delphi-based semistructured interview with a structured questionnaire is the methodology. A generic questionnaire has been developed with list of questions. The questions are asked with all the employees in the organization. The subcomponents of three *KM* pillars such as people, process, and technology have been derived and categorized based on semistructured interview. For an improved clarity, categorizations of computable characteristic of the three above-mentioned pillars are arrived through a semistructured interview conducted with 43 executives belonging to 32 manufacturing organizations. Based on the semistructured interview and Delphi study, the subcomponents of three *KM* pillars toward the organizational readiness are identified at the end of the third round of Delphi study. The critical subcomponents of *KM* pillars are people – skills, leadership, culture/structure, and exploitation; process – processes, measures, explicit knowledge, and tacit knowledge; and technology – knowledge centers and infrastructure. The sample of semistructured interview is shown in Table 3.1. Once the subcomponents are derived based on the semistructured interview and Delphi study, the readiness assessment framework should be developed by involving all the three *KM* pillars and subcomponents of *KM* pillars. The readiness assessment framework with all the *KM* pillars and subcomponents of *KM* pillars should be derived in this step, and the derived framework based on 40 executives is depicted in Fig. 3.4:

People: skills, leadership, culture/structure, and exploitation – these address the "mindset" and relate to attributes of assessing community and civilization. People element also includes customers.

Process: processes, measures, explicit knowledge, and tacit knowledge – these are the facilitators for people to strap up the knowledge in a standardized way across the organization.

Technology: knowledge centers and infrastructure – these address the enablers and facilities which help people and process to bind the utmost out of the *KM* initiative.

To deal with the needs for an organization's *KM* readiness, one needs to consider that it is time dependent and would be pretentious with any change in the basic subcomponents of the three critical pillars – people, process, and technology. Hence,

Table 3.1 Sample template for semistructured interview and Delphi study

No.	Knowledge components	Objective	Function/CSO	Gap	Reason for gap		Solution for the gaps
					Availability	Accessibility	
1	Cost of materials, man, and machine	Functional	Finance and accounts	Currently no structured data base(DB) is available that can provide such information	X	X	Database – cost
2	Latest industry trends in accounting and costing	Functional	Finance and accounts	Currently no structured data base is available that can provide such information	X	X	Industry analysis
3	Machine price data, application details, operations data, ratios	Functional	Finance and accounts	Data, though available, is not captured or stored in a structured manner		X	Product data base
4	Order to cash process, procure to pay process	Functional	Finance and accounts	Processes not documented properly and not stored in a structured manner	X		Standard operating procedures (SOPs)
5	Finance act(s), state legislations	Functional	Finance and accounts	Not easily accessible		X	Document library
6	Classification of accounts – capital and revenue, income and expenditure, taxation data (central excise, sales tax, income tax), TAX reporter	Functional	Finance and accounts	Data, though available, is not captured or stored in a structured manner		X	Records library
7	Management and technical institutes	Functional	Finance and accounts	Key contacts not available	X		Database – institutes
8	Technical newsletter/business/economy-related articles/analysis/surveys/journals/periodicals/annual reports	Functional	Finance and accounts	Single-point access to all resources not available	X	X	Technical information center

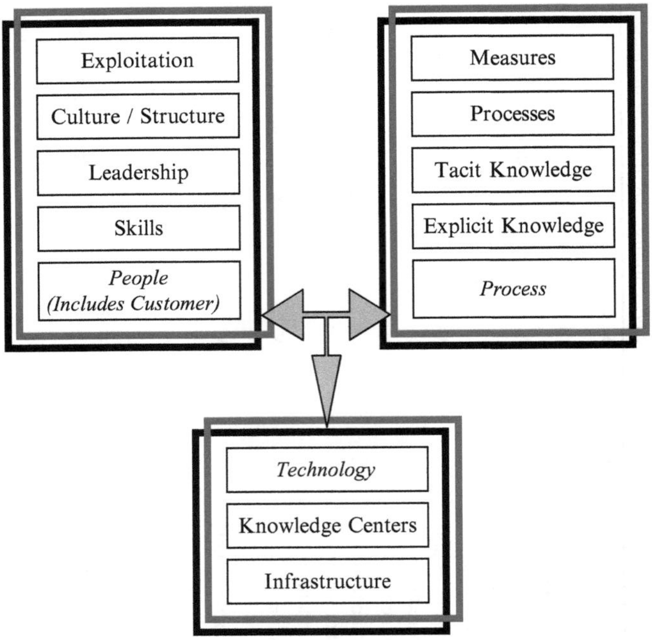

Fig. 3.4 Readiness assessment framework with *KM* pillars and subcomponents

the subcomponents under each pillar must be viewed in lieu with context of organization, and a suitable set of subcomponents under each pillar would need to be defined for different organizations and also for different readiness levels. Any manufacturing organization can use this procedure to develop the readiness assessment framework. This generic readiness framework developed in this research can be used as a base for any manufacturing organization.

In Step 1 of Phase 2, a weightage approach for the questions is used as a research methodology. Weightage should be provided to each question, and based on the weightage of the question and also based on the number of participants/executives participated in the survey, the weightage proportion or percentage of subcomponents of *KM* pillars for the organizational readiness needs to be identified. The survey involves ten questions that aimed at providing a quick check of where an organization comes along ten subcomponents under three critical success pillars. The scores are arrived by calculating the mean and standard deviation of all response weightages. The weightage of each subcomponent is calculated as follows:

$$\text{Weightage for each component} = \left(\frac{\text{Summation of the weightage of each component for all responses}}{\text{Summation of maximum scale of each component for all responses}} \right)$$

The weightage proportion for all the subcomponents of *KM* pillars are calculated based on the 40 executives and the benchmark values are detailed in Fig. 3.5.

The research design for readiness assessment is detailed in Fig. 3.6.

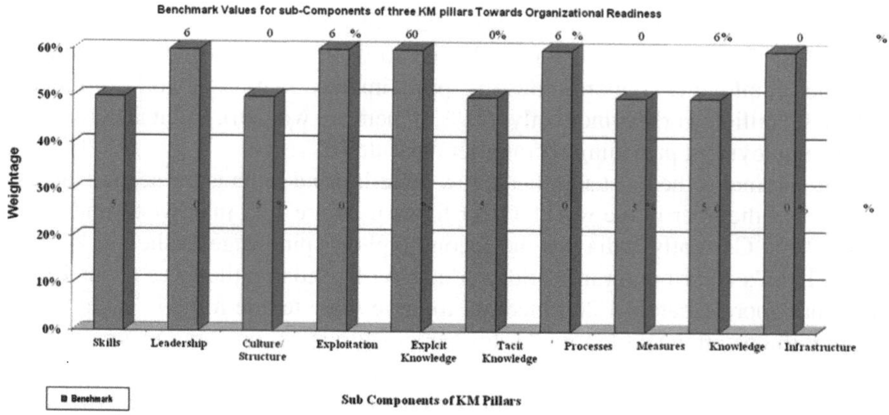

Fig. 3.5 Benchmark values for subcomponents based on questionnaire survey

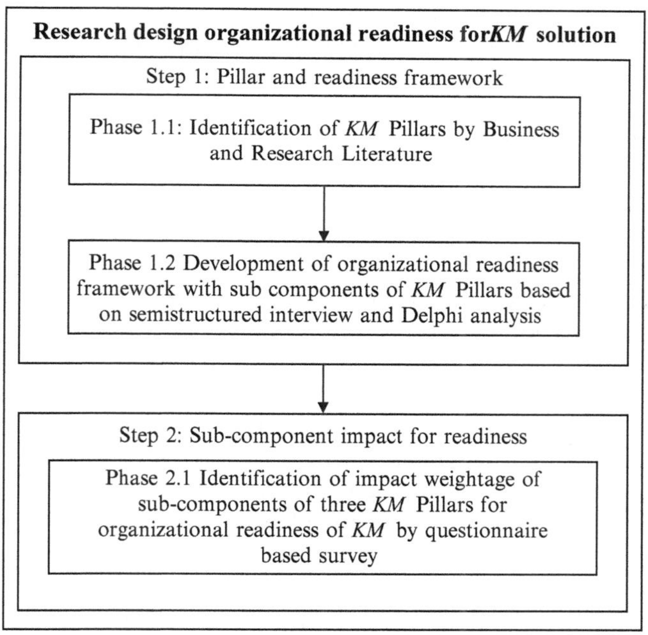

Fig. 3.6 Research design

3.4 Case Study Demonstration

The entire research process and methodology is demonstrated and applied through a real-life case study for Indian textile machinery manufacturing company. The Indian textile industry is the second largest in the world. Indian textiles also account for 38 % of the country's total exports of nearly $45 billion and are, therefore, a very important industry. The textile industry is the single largest foreign exchange earner for India. Currently, it accounts for about 8 % of gross domestic product (GDP), 20 % of the industrial production, and over 30 % of export earnings of India. It generates employment opportunities for approximately 38 million workers directly and 54.85 million workers indirectly (60 % of them are women), and it is the second largest employment providing sector after agriculture.

Cotton remains the most significant raw material, and India is the second largest producer of the fiber in the world. Other fibers used are silk, jute, wool, and man-made fibers. Currently India has the second highest spindleage in the world after China. India's contribution in world production of cotton textiles was about 15 %. There are approximately 1,200 medium to large scale textile mills in India. India has 34 million cotton textile spindles for manufacturing cotton yarn. Approximately 120 companies manufacture the complete range of textile machinery. India has 3 % share in the export production of clothing. Gujarat and Tamil Nadu are the two largest textile manufacturing states of India. The USA is known to be the largest purchaser of Indian textiles. Also India has a marked presence in the United Arab Emirates (UAE), Saudi Arabia, Canada, Bangladesh, China, Turkey, and Japan.

Textile machinery is used in the fabrication and processing of fabrics, textiles, and other woven and nonwoven materials. The major product segments under the head "textile machinery" include textile processing machinery and textile working machinery. A further classification of these two segments may be fiber–fabric machinery (cleaning and opening machinery, carding and combing machinery, drawing and rowing frames, spinning and twisting frames, yarn winding machines, yarn preparing machines, and other fiber-to-fabric machines), fabric machinery (weaving machinery, knitting machinery, and other fabric machinery), other textile machinery (bleaching, mercerizing and dyeing machinery, textile printing machinery, textile finishing machinery, and other complete textile machinery), and textile machinery parts and accessories.

The case organization established in 1962 is currently one of the three global companies to manufacture the entire range of textile machinery and is the topmost manufacturer of textile machinery in India. The case organization has a market share of around 60 % in the spinning machinery textile industry in India and located in South India, the state of Tamil Nadu. The case organization was the first Indian company to introduce the automatic bale plucking machine which surpassed any other bale plucking machine in the world. There are at least 20 domestic companies offering textile machinery in India, and the major suppliers other than the case organization are located in northern regions. Approximately USD80 million in Indian textile machinery is exported to other developing countries. World production of manufactured textile fiber is projected to rise over 62 million metric tons in 2012. Overall growth rise is because of demand for textile fibers used in upholstery, household furnishings and apparel, and floor coverings. The recent development is also in the field of medical textiles, geotextiles, agrotextiles, and protective textiles. This case organization decided to implement the *KM* solution to improve the productivity and to enhance the position of organization. The case organization is traditional and the readiness for any new change is a great challenge. So, there is a need to understand the readiness of the organization toward the initiative of implementation of *KM* solution. The readiness assessment model is devised for this case organization.

For the implementation of *KM* solution in the case study organization, we have considered three *KM* pillars as stated earlier. The *KM* pillars for this organization were identified based on business and research literature. The subcomponents of *KM* pillars are identified, and the readiness assessment framework is developed based on semistructured interview and Delphi study. The weightage for the subcomponents of *KM* pillars are identified based on questionnaire survey. The respondents include various hierarchical positions of employees like members of the management committee/board of organization, few strategic top level executives, and few tactical and operational employees. Thus, 216 executives of this textile machinery manufacturing organization were interviewed for the study.

The details of each phase of research process and methodology are explained for the case study.

Step 1 of Phase 1

The same *KM* pillars such as people, process, and technology are leveraged for this case study based on business and research literature.

Step 2 of Phase 1

The subcomponents are identified based on semistructured interview and Delphi analysis. A total of 216 executives participated in this Delphi study. We conducted three rounds to finalize the outcome of the subcomponents. All three rounds were conducted in a big auditorium for an entire day. The 216 executives included key members from the bottom to the top level of the organization, who are considered as a sample size of the 700 employees in the organization. The sampling is done on a random basis. The questions are rated, and based on the outcome of the third round, the same subcomponents indicated in Sect. 3.3 are identified. The readiness assessment framework for the case study is indicated in Fig. 3.7.

Step 1 of Phase 2

The impact weightage of the subcomponents are identified based on questionnaire survey, and impact weightage is detailed in Fig. 3.8.

The ranks are converted to relative percentage by dividing each rank by the total of all ranks for the group of measures/characteristic. This approach is similar to the method used in Pareto analysis wherein problem frequencies are converted to percentages to show relative performance. The percentages better highlight differences in the importance of the characteristic. The assessment outcome is purely based on individual perception of how knowledge is being managed in their respective areas and should be used for planning and implementing *KM* at an organization. This assessment is also only an indication of where organization stands as of today in terms of the critical pillars that contribute to the *KM*. The comparison of case study values against the benchmark values is depicted in Fig. 3.9.

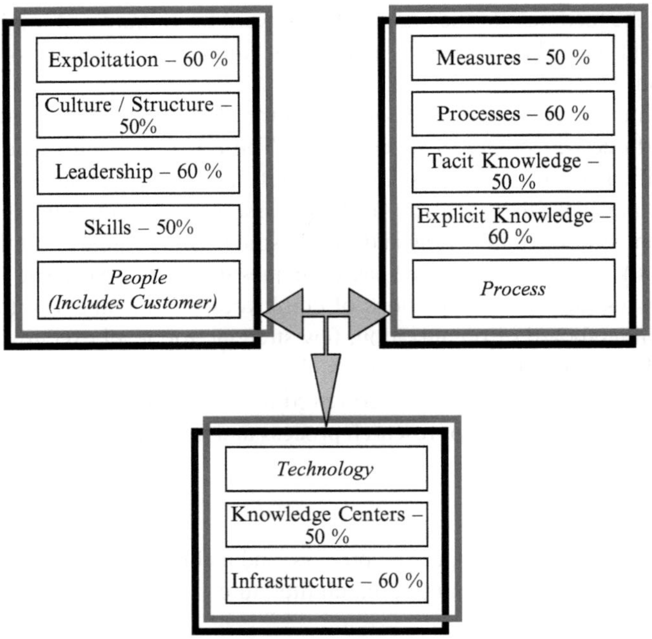

Fig. 3.7 Readiness assessment framework for the case study

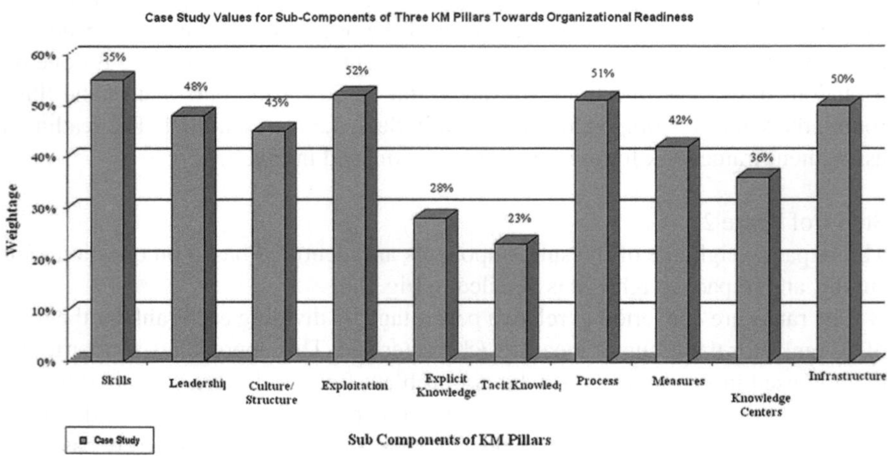

Fig. 3.8 Weightage for the case study

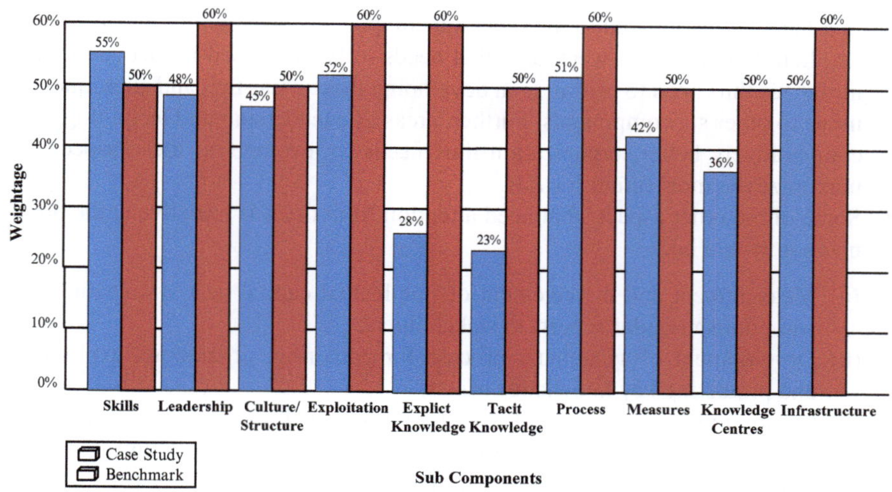

Fig. 3.9 Comparison of case study and benchmark values

3.5 Managerial Implications

The outcome of the readiness assessment for textile machinery manufacturing organization is:

- Organization has a compelling knowledge vision and strategy, actively promoted by the top management that clearly articulates how KM contributes to achieving organizational objectives. This is highly evident from the fact that the assessment outcome shows around 50–60 % for leadership.
- Another aspect that is highlighted in the assessment outcome is that the organization has a fairly good technological infrastructure to support an initiative like *KM*. Even in the existing scenario, important information can be found/shared on the intranet.
- Organization also portrays a fairly good level of exploitation of knowledge. It is being seen that knowledge and experiences are converted into projects or initiatives that help the organization's growth.
- The primary obstacle to *KM* in the organization comes from explicit knowledge and tacit knowledge. These are pretty low in the organization and thus shows that the organization does not maintain any knowledge inventory and also that there is no clear ownership of knowledge entities that is readily accessible across the organization. Further, though the organization has a rich pool of domain experts with key knowledge, there is no mechanism in place to codify, capture, and use this for the organization.
- One aspect that would need concentration on is the knowledge centers to coordinate knowledge repositories and act as focal point for provision of information to support key decision making and business functions. It is expected that the proposed *KM* cell will address this aspect.

- In a nut shell, compared to the other subcomponents, availability of knowledge (explicit and tacit) is a weak area that needs to be addressed immediately. *KM* initiatives thus need to prioritize in developing these areas first and subsequently move to other subcomponents. Further, areas like leadership and technology are comparatively better; organization thus needs to leverage on these aspects to maximize the momentum for *KM*.
- Some of the other aspects that need attention from the *KM* initiative at the organization would be:

  (a) Measurement and management of intellectual capital in a systematic way, and publish regular reports to stakeholders.
  (b) Development of systematic process for gathering, organizing, exploiting, and protecting key knowledge assets.
  (c) Creation of culture of knowledge sharing across departmental boundaries.
  (d) Identification and assignment of specific knowledge roles, and ensure that all senior managers and professionals are trained in *KM* techniques.

- Over all there is a positive attitude toward the *KM* initiative across the organization with a lot of expectations. The same kind of enthusiasm needs to be sustained by ensuring that the expectation is met to the maximum possible extent from the implementation of *KM*.

3.6 Conclusions

KM is all about the ability of organizations to leverage the intellectual assets quickly and accurately. To achieve successful *KM*, a readiness assessment approach is vital to investigate an organization's knowledge "health." The readiness assessment approach provides an evidence-based assessment of where the organization needs to focus before the implementation of *KM* effort. It can reveal the organization's needs, strengths, weaknesses, opportunities, threats, and risks toward the implementation of *KM*. As many methodologies of readiness assessment approach suggested in most of the previous research were very general and aimed at company-wide uses, it may not be appropriate for companies as a generic format. In this research, a systematic readiness assessment approach is proposed. The readiness assessment approach will address the organization's *KM* awareness level; analyze its knowledge support processes, its structures and roles; and identify the key business areas that serve as the targets for the KM initiatives. The readiness assessment approach helps the organization to develop *KM* strategies that linked to its business strategies. The organization will be able to set the criteria for choosing the appropriate knowledge that it planned to pursue and formulate plans to capture and share it. Also the organization will be able to plan effectively for future *KM* activities. It successfully develops a detailed *KM* project plan toward the organization readiness to reserve appropriate resources and manpower for project implementation and monitor the progress of various *KM* activities. *KM* is still in its infancy in India. Very few companies have appointed dedicated personnel to take responsibility of *KM*. In most firms *KM* has been tagged on to somebody's existing responsibilities, often resulting in a step-motherly treatment. But this situation cannot last given the increasing competitive business environment in India. *KM* is no longer a luxury for Indian companies. It is a necessity that can make all the difference between survival and an early demise.

Chapter 4
Behavior Assessment for Knowledge Management Solution Implementation

4.1 Introduction

Among the three strategic components of *KM*, i.e., people, processes, and technology, the most imperative one is "people." *KM* essentially depends on the readiness and behavior of individuals to indicate the ownership, contribution, use, and reuse of knowledge and to share it when requested. Processes and technology can aid to enable and assist *KM*, but it supports in creating, contributing, sharing, using, and reusing knowledge for and by the people and this is the fundamental principle of *KM*. Behavior of people is a key component which supports to achieve organizational objectives (Krishnan 2001). Good and right behavior of people enhances firms to create and increase the utilization of their institutional or collective knowledge.

In the earlier days of *KM*, many organizations focused primarily on processes and technology, which saw futile ends. Having made huge investments in the latest systems, organizations realized that people are reluctant to create, contribute, and share knowledge. It was identified that people simply did not use/reuse the knowledge that was created and stored. This leads the *KM* system as the "*KM* graveyard." Organizations today have learned that it is the people who "make or break" *KM* initiative. It is therefore essential to create conducive organizational culture to facilitate the knowledge sharing and utilization. The culture of an organization is basically the set of values, beliefs, assumptions, and attitudes that are deeply held "by the people" and "for the people" in an organization. They influence the decisions that people take and also the ways in which employees behave. Organizational cultures run deep; the older and the bigger the organization, the more deep-rooted they will be. Thus, one may observe that organizational culture and individual behavior are two aspects of *KM* which are inextricably linked. People in any organization are influenced by the culture of their working environment. The organizational culture is a direct reflection of its own modal organizational personality, the degree of homogeneity, and the strength of a particular orientation in that organization that allows smooth management. It resulted from four factors. They are (1) socialization, (2) selection process to ensure some level of personality and homogeneity,

K. Ganesh et al., *Design and Development of Knowledge Management for Manufacturing: Framework, Solution and Strategy*, DOI 10.1007/978-3-319-02892-7_4,

(3) rewards in the organizations to selectively reinforce some behavior, and (4) attitudes and the promotion decision usually accounting performance as well as personality. So it becomes very important at the onset of the *KM* journey; the values, attitudes, and behavior of the people in the organization should be assessed that constitute barriers to seeking, sharing, and using knowledge.

4.2 Research Gap Based on Literature

The need for having a holistic view about the structure, tasks, and people in organizational transformation is highlighted in literature. It is also clearly evident that employees are the key for any organizational change like *KM*. From the detailed literature survey, the research gap is shown in Fig. 4.1.

Identification of behavior types of employees in an organization, identification of proportion or percentage of existence of behavior types in an organization, and setting behavior targets for improvement are not widely explored in the literature. This behavior assessment is critically important in an organization before the implementation of *KM* solution.

Based on the behavior assessment, organization should initiate projects with behavior targets. Otherwise, implementation of *KM* solution will not be successful for any organization. The primary intention of behavior assessment is to assess the exact behavior pattern among the people in the organization toward the implementation of *KM* solution. The objective of this module is to design a generic conceptual

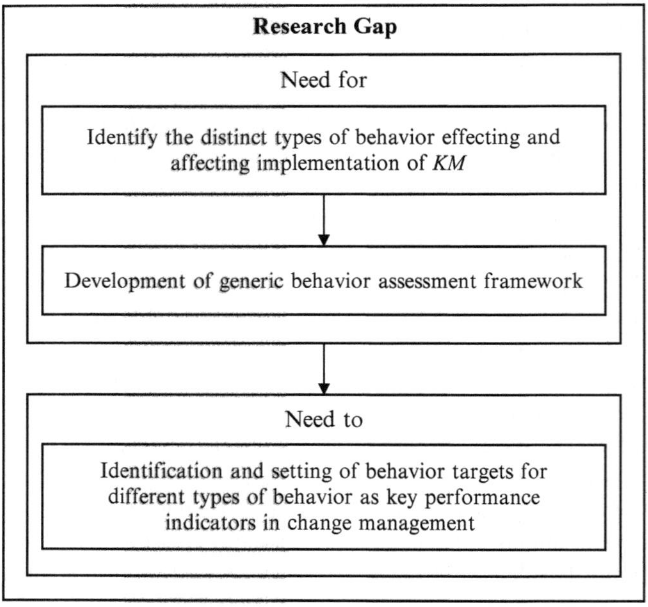

Fig. 4.1 Research gap

framework and generic solution procedure for behavior assessment for any manufacturing organization with a set of behavior types and targets. The types and targets are derived from the literature, and those can be changed with respect to the mission and vision of the organization.

4.3 Research Process and Methodology

The need for profiling employees and their resultant behavior for a *KM* initiative is detailed through this study. The behavior assessment framework has been developed with a set of behavior targets.

The research process is divided into two phases:

Phase 1: Identification of behavior types and behavior assessment.

Step 1 of Phase 1: Identification of distinct types of behavior in an organization through semistructured interview with a structured questionnaire.

Step 2 of Phase 1: Assessing the proportion of percentage of behavior types available in the organization.

Phase 2: Based on the behavior assessment, the behavior types are set to achieve the positive aspects of behavior in the organization.

This methodology is designed in such a way that any organization can use this as a base and they can identify various behavior types available in the organization to assess the proportion or percentage of behavior types available in the organization and to set the behavior targets in order to improve the positive behavior in the organization for the implementation of *KM*.

The research methodology used in the research process is detailed here:

In the Step 1 of Phase 1, a Delphi-based semistructured interview with a structured questionnaire is the methodology. A generic questionnaire has been developed with a list of questions. The questions should be asked with all the employees in the organization. Then the answers of the questions need to be linked with the behavior types. Out of five questions for each behavior type, if more than three questions are linked through "yes" category, then that behavior is considered for that organization. Likewise, the available behavior types in the organization should be determined.

The "people factor" is responsible for the success or failure of any *KM* initiative in any organization that can be a resultant of distinct employee behavior types in the organization. Based on the research and business literature, there are six types of behavior suggested for Step 1 of Phase 1. The structured questions designed all revolve around these six types. The semistructured interviews are conducted with 43 executives belonging to 32 manufacturing organizations. The structured questions and the linkage of questions to the behavior types can be shared on the request basis from the authors of this research paper. These six types can be used as a base for any manufacturing organization. Almost all the behavior types are captured in these six types and detailed below:

- *Skeptic*: People who do not believe things they are told or as others might believe. The person is of investigative nature (positive) on one hand and suspicious (negative) on the other.
- *Convert*: People who tend to accept an idea readily and are able to align themselves easily.
- *Cynic*: People who make things seem worse than they actually are and by nature they tend to denigrate or belittle everything.
- *Procrastinator*: People who try to postpone things by excuses of doing something more important and would wait and wait until the mood strikes him/her that it is time to do the work.
- *Potential*: People who see the importance of an idea but need some guidance and motivation to actually participate wholeheartedly. These people are normally the creative, innovative trendsetters of the organization.
- *Rebel*: People who love to be contrary and point out others' flaws unnecessarily. They tend to value their own opinion above all else.

In Step 2 of Phase 1, a weightage approach for the questions is used as a research methodology. In Step 2, weightage should be provided to each question, and based on the weightage of the question and also based on the number of points linked to each behavior type, the proportion or percentage of behavior types in the organization needs to be identified. In Step 2 of Phase 1, the questions need to be weighed based on the preference of top management of organization in the scale of 1–10. Then the number of questions linked by the total number of employees in the organization to each behavior type should be identified.

If question 1 is "yes" and answered as "yes" by 50 out of 100 employees and question 1 is linked to one of behavior type 1, then behavior type 1 is rated as 50 % for question 1. Then if question 1 is ranked as 6 out of 10 scale, then the score for the behavior type 1 is calculated as follows: $= 50\ \% * 0.6 = 30\ \%$.

Likewise this should be done for all the questions related to each behavior type. If the total number of questions linked for one type of behavior is 5, then the average figure should be considered for that behavior type. In Phase 2, after assessing the proportion or percentage of behavior type in the organization, the respective behavior types should be set for the employee and the management. Proper team should be designed by the management to execute the behavior targets. Separate project charters should be designed for each behavior target. Performance monitoring system and communication plan and channel should be set to execute and control the behavior targets effectively.

The behavior targets for the employees and management are derived based on the business and research literature and also based on the interviews with 43 executives belonging to 32 manufacturing organizations. This can be used as a basis for any manufacturing organization and is detailed in Table 4.1. Overall, 43 executives participated in the study, from different organizations varying in size from 200 to over 20,000 employees. Interviews were conducted in person and also over the telephone. All interviews were conducted based on the interview protocol. After each interview, notes were reviewed to identify potential challenges or problems.

The diagrammatic representation of research design is detailed in Fig. 4.2.

Table 4.1 Types of behavior with behavior targets

Types	Behavior targets	
	Employee	Management
Convert	Actively participate in the mentoring culture of the organization	Encourage peer-to-peer discussions and mentoring practices. Assign them a coach or mentor to help them with development
	Self-motivation	Issue proper guidelines and instructions
		Treat him as an internal customer and the new idea should be "sold" to him
Cynic	Avoid unrealistic expectations by having proper knowledge of policies of the company	Manage expectations: must be careful about the promises they make or are perceived to make. High expectations can encourage engagement and hard work, but unrealistic expectations often result in disillusionment and situational cynicism. Low expectations, on the other hand, seem to be a primary cause of ideological cynicism
	Manage conflict and negative emotions	Proper communication: secrets and surprises breed cynicism. Cynics hide inferiority complex in cynicism. Hence, it may be a good idea to give them ownership to make things happen
	Ask the right questions about new initiatives and organization's strategies	Focus on middle managers: because of the extent of interaction with the shop floor people, this group should ideally be groomed as knowledge engineers
		Peer-to-peer discussion: in carefully structured small-group discussions, coworkers can challenge the perceptions and assumptions of the ideological cynics. Peers should be selected carefully and should avoid putting cynics together
		Involve the employees in the decision-making process whenever possible. This takes forward the earlier point on ownership
Procrastinator	Effective time management	Map current time management skills of employees and techniques to avoid interruptions
	Get into the "do it now" habit	Workshops on motivation aspects in workplace and beyond. Workshops to focus on time management, prioritization (LN itself can be used as time management tool)
	Eliminate indecision and hesitation	Clarification on expectations from the employees
	Cultivate the ability to prioritize and say NO	Case studies on time management goof-ups can be presented and discussed

(continued)

Table 4.1 (continued)

	Behavior targets	
Types	Employee	Management
Potential	Ask the right questions	Create a work environment where people enjoy what they do, feel like they have a purpose, have pride in what they do, and can reach their potential
		Foster creativity
		Showcase case studies about successful implementation of *KM*
		Give ownership and timelines
Rebel	Manage frustrations and disappointments	Transparency in management
		Training on team building
		Clearly defined incentive plans and rewards and recognition policies
		Provide employees with channels of communication to report supervisory conduct and decisions they regard as discriminatory, harassing, or otherwise wrongful. Promptly investigate these complaints and seek legal advice for any that remain unresolved
		Make sure that morale of the other workers does not suffer because of a disruptive employee
Skeptic	Ask the right questions about new initiatives and organization's strategies	Showcase case studies about successful implementation of *KM*
		Conduct open house with the employees to drive out any kind of skepticism in their minds. Formal record and preparation of action plan on the ideas and suggestions made by employees
		Periodic communication plan on targets and actual
		Explain the "big picture" for the company and how this influences employee's employment and growth
		Make the employee contribute one or two ideas and try to get buy in of the "complete picture"

4.4 Case Study Demonstration

The entire research process and methodology is demonstrated and applied through a real-life case study for Indian textile machinery manufacturing company.

The company is a global player and one among the three manufactures of the end-to-end range of textile machinery, i.e., from blow room machinery to ring

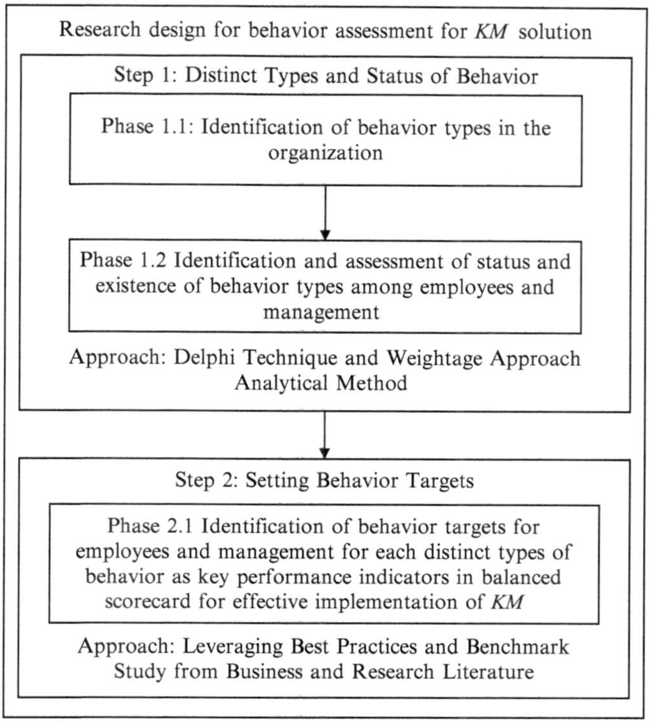

Fig. 4.2 Research design

frame. The company has 60 % market share in the domestic textile spinning machinery industry. The "state-of-the-art" foundry makes precision castings for industries the world over. It is the only company in Asia outside Europe to manufacture original equipment products for MIKRON of Switzerland. The company has won the top export award in textile machine exports for the past 7 years. The company has exclusive independent business units for all its products in the complete spinning system. Each of the following such as blow rooms and all preparatory machines such as carding, combing, drawing, speed (longitudinal), and ring frames products has independent business units.

For the implementation of *KM* solution in the case study organization, we have considered six distinct types of employee behavior as stated earlier. The behavior types in the organization were identified with the use of a Delphi/expert opinion technique and were measured using a semistructured interview along with a structured questionnaire. The respondents include various hierarchical positions of employees like members of the management committee/board of organization, few strategic top level executives, and few tactical and operational employees. Thus, 216 executives of this textile machinery manufacturing organization were interviewed for the study.

The details of each phase of research process and methodology are explained for the case study.

Fig. 4.3 Behavior survey assessment outcome

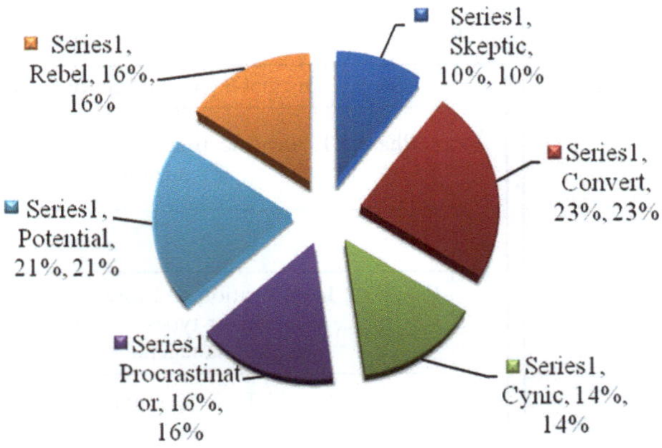

Step 1 of Phase 1
The designed questionnaire is used for this study. A total of 216 executives participated in this Delphi study. We conducted four rounds to finalize the outcome of the structured questionnaire. All four rounds were conducted in a big auditorium for an entire day. The 216 executives include key members from the bottom to the top level of the organization, who are considered as the sample size of the 700 employees in the organization. The sampling is done on a random basis. The questions are rated and based on the outcome of the 4th round and the behavior types available in the organization are identified. We found that all six types of behavior were found in the people of the organization.

Step 2 of Phase 1
The questions are asked to rank all the employees. Then the assigned rank is considered for the questions. Each question is answered by all the employees. Then, we calculated the number of questions answered in the "yes" category. The number of "yes" is counted for each question, and the percentage is calculated against the total number of employees. This is done for each question, and then the percentage is multiplied by the rank of the questions. Then the question are grouped and linked to each behavior types. Then the average of all questions for each behavior type is calculated, and that is the proportion or percentage of each behavior type existing in the organization. A percentage analysis has shown us how the employees of this case study organization can be profiled. And the graph is shown in Fig. 4.3.

Behavior assessment was carried out for the case organization, and it is suggested that there are six behavior types in the organization, namely, skeptic (10 %), convert (23 %), cynic (14 %), procrastinator (16 %), potential (21 %), and rebel (16 %), and the targets for certain behavior types need to be reviewed and revised for the improvement toward the acceptance for *KM* solution implementation.

Phase 2
Then after identifying the percentage of behavior type, the behavior targets for employees and management are devised based on the available behavior target framework. The behavior targets for employees and management strategies for each distinct types of behavior are devised and tabulated in the Table 4.2.

Table 4.2 Behavior targets for case study

Types	Behavior targets Employee	Management
Convert	Mentoring training for every 1 month	A mentor is assigned to a set of 10 employees apart from the supervisor or manager
	Training on self-motivation	Provide booklet on guidelines and instructions
		Develop an internal customer team and assign all the "convert" employees to this team
Cynic	Training on policies of the company	Set realistic KPIs for this employees and reduce number of KPIs for this employees
	Training on conflict and negative emotions	Assign reporting team to this employees
	Information on new initiatives and organization's strategies	Focus on middle managers: because of the extent of interaction with the shop floor people, this group should ideally be groomed as knowledge engineers
		Assign as lead for peer-to-peer discussion
		Increase the decision-making ownership for these employees
Procrastinator	Training on effective time management	Map current time management skills of employees
	Get into the "do it now" habit	Workshops on motivation aspects in workplace, time management, prioritization
	Training on team work	Get expectations of these employees often
	Training on prioritization	Give ownership for presentations on time management
Potential	Training on right questions probing	Create a entertainment environment in the work
		Set creative tasks
		Training on *KM* success
		Give ownership and timelines
Rebel	Training on managing frustrations and disappointments	Transparency by management should be increased for these employees
		Training on team building
		Incentive plans and rewards and recognition policies should be designed
		Create complaint resolution team
		Create moral team
Skeptic	Training on new initiatives and Organization's strategies	Training on *KM* success
		Conduct open house with the employees
		Periodic communication plan on targets and actual
		Explain the "big picture" for the company and how this influences employee's employment and growth
		Make the employee contribute one or two ideas and try to get buy in of the "complete picture"

4.4.1 Interpretations of Behavior Assessment of the Case Study

The interpretations of the behavior survey assessment of the case study are detailed below:

- It is clearly evident from the chart above that case study organization primarily has a convert and potential behavior type.
- This is highly encouraging and shows that there is a positive environment in case study organization for *KM* since convert and potential are key drivers for the success of *KM* in any organization.
- However, though not highlighted, there is a significant level of rebel, procrastinator, and cynic behavior types.
- This is a reason for caution as the current behavior trend can shift even to the negative environment if the right steps are not taken to tackle the different behavior types.
- The subsequent section will outline the behavior targets for both the employees and the management for all the five predominant behavior types among the employees in the case study organization.

4.5 Managerial Implications

Nelson and Quick (1994) have found that the individuals can be divided into disengaged, disidentified, disenchanted, and disoriented according to their reactions to change. Disengagement is psychological withdrawal from change. The employee may appear to lose initiative and interest in the job. Employees who disengage may fear the change but take on the approach of doing nothing and simply hoping for the best. Disengaged employees are physically present but mentally absent. Disenchantment is also a common reaction to change. It is usually expressed as negativity or anger. Disenchanted employees realize that the past is gone, and they are mad about it. They may try to enlist the support of other employees by forming coalitions. Disoriented employees are lost and confused, and often they are unsure of their feelings. They waste energy trying to figure out what to do instead of how to do things. Disoriented individuals ask a lot of questions and become very detail oriented. They may appear to need a good deal of guidance and may leave their work undone until all of their questions have been answered. "Analysis paralysis" is characteristic of disoriented employees. Another reaction to change is disidentification. Individuals reacting in this way feel that their identity has been threatened by the change, and they feel very vulnerable. Many times they cling to a past procedure because they had a sense of mastery over it, and it gave them a sense of security. The findings of this study can be related to the profiling of Nelson and Quick (1994). DeMeuse and McDaris (1994) have profiled individual reactions to change using R-T-C (reaction to change) questionnaire as "change agents," "change compliers," and "change challengers." The profiling of the present study also goes hand in hand with these reaction dimensions.

The present results of the manufacturing industry holds good that in India there is a fare chance of having positive environment to change. Based on the literature review, the fundamental emotional reasons why employees resist *KM* could be that:

- It is not timely
- Lack of trust on the management
- Feeling of losing one's power as an expert
- Feelings of doubt
- Insufficiency

Lewin (1939) claims that for any organization to manage a change effectively, three different approaches should be targeted. They are to educate the employees about the inevitable need that follows through rational empirical approach, normative educative approach, and finally power coercive approach.

Thus, the climate setting for change can happen through a series of systematic well-defined steps, like:

Step 1: Creating strategic vision and mission for the employees.
Step 2: Setting behavior targets.
In this step, the behavior targets devised based on the best practices and benchmark study using Supply Chain Operations Reference (SCOR) Model and other relevant resources. Identification of behavior targets for employees and management for each distinct types of behavior as key performance indicators is through the balanced scorecard for effective implementation of *KM*.
Step 3: Use these targets for performance appraisals.
Step 4: Reward mechanisms.
Step 5: Maintaining the climate.

The essence of management strategies can be summarized in a 5-step *PRIDE* model suggested as follows:

- **P**rovide a positive working environment.
- **R**ecognize, reinforce, and reward everyone's efforts.
- **I**nvolve everyone.
- **D**evelop skills and potential.
- **E**valuate and measure continuously.

These steps would ensure contented employees who would work toward for a better future promising better productivity in any turbulent changes.

4.6 Summary

KM is often used to improve performance through enhancing business processes, believing that even if immediate measurable paybacks are few, performance improvements will reap benefits over the long term. Research in *KM* has concentrated, with few exceptions, on organizations with a future, assuming that changes today will over time feed through to performance improvements. The behavior assessment model is a generalized model used to assess the behavior types in the organization with the devised framework of behavior targets. It also set the behavior targets for employees and management toward the effective implementation of *KM*. This research connects the gap between industry and academia by providing methods and tools to increase knowledge work productivity through development of a structured *KM* initiation method and identifying impeding factors for its successful implementation. This study has contributed to the literature by bringing out an interdisciplinary outlook and practical underpinnings for understanding *KM* initiation through applied behavioral psychology. On a careful examination, there are a number of limitations affecting the scope of these findings. For one, our survey respondents were all from manufacturing sector. Future studies should be extrapolating this research theme to other industry types. Secondly, interviews and surveys were confined to a sample from manufacturing industry. A larger sample size would add more statistical significance to the results. Future research would also benefit from wider diversity within the sample, in terms of firm size, industry, and geographic regions. Given these limitations, however, there are still lessons that can be drawn from this study. One major key point is that this study holds the basis for profiling of the employees well in advance to find the possible change agents and to suggest strategies for effective strategic management that would prompt *KM* initiation and organizational transformation both positively and productively.

Chapter 5
Taxonomy and Technology Architecture for Knowledge Management Solution Implementation

5.1 Introduction

Enterprises are realizing how important it is to "know what they know" and be able to make maximum use of the knowledge. This knowledge resides in many different places such as databases, knowledge bases, filing cabinets, and peoples' minds and is distributed right across the enterprise. All too often, one part of an enterprise repeats work of another part simply because it is impossible to keep track of, and make use of, knowledge in other parts. Enterprises need to know:

- What their knowledge assets are
- How to manage and make use of these assets to get maximum return

Most traditional company policies and controls focus on the tangible assets of the company and leave unmanaged their important knowledge assets. Success in an increasingly competitive marketplace depends critically on the quality of knowledge which organizations apply to their key business processes. For example, the supply chain depends on knowledge of diverse areas including raw materials, planning, manufacturing, and distribution. Likewise, product development requires knowledge of consumer requirements, new science, new technology, marketing, etc. Philosophically, "Knowledge is experience. Everything else is just information" – Albert Einstein. "The wise see Knowledge and Action as one" – The Bhagavad Gita. *KM* is not only about managing these knowledge assets but managing the processes that act upon the assets. These processes include developing, preserving, using, and sharing the knowledge. By all accounts, *KM* is the ability to create and use knowledge to achieve organizational objectives (Krishnan 2001). *KM* is a business process through which firms create and use their institutional or collective knowledge.

Organizational learning is important for a firm's productivity growth. It is found that information and communication technology alone do not support organizational learning and therefore do not enhance productivity. This research has demonstrated that in the case of manufacturing companies, some computer-based systems do support organizational learning. Taxonomy is the apex operational structure of

K. Ganesh et al., *Design and Development of Knowledge Management for Manufacturing: Framework, Solution and Strategy*, DOI 10.1007/978-3-319-02892-7_5,

the enterprise, and it covers and categorizes all functional aspects of the enterprise under different categorizes. Taxonomy is a standardized set of terms, hierarchically organized, used to categorize information and knowledge. The taxonomy generally reflects how organizations think about the business, how the firms organize to conduct business, and/or how and what the firms deliver to their customers. The hierarchical organization is a useful way to display relationships among terms and makes it easier to find like items at more general or more specific levels. At its most basic level, the taxonomy standardizes what we call things, making a consistent connection between an idea or concept and the words we use to describe it. This standardization makes it easier for the ultimate user to find what he or she is looking for. Based on semistructured interview, expert's opinion, and Delphi study, we propose a generic framework of taxonomy architecture for a typical manufacturing organization in this chapter. The taxonomy should also be extensible to address non-document form of outputs as well. An organizational structure is mainly a hierarchical concept of subordination of entities that collaborate and contribute to serve one common aim.

Taxonomy is a standardized set of terms, hierarchically organized, used to categorize information and knowledge. Taxonomy generally reflects how we think about our business, how we organize ourselves to conduct business, and/or how and what we deliver to our customers. The hierarchical organization is a useful way to display relationships among terms and makes it easier to find like items at more general or more specific levels. At its most basic level, the taxonomy standardizes what we call things, making a consistent connection between an idea or concept and the words we use to describe it. This standardization makes it easier for the ultimate user to find what he or she is looking for. In other words, taxonomy is the apex operational structure of the enterprise, and it covers and categorizes all functional aspects of the enterprise under different categories. The taxonomy should also be extensible to address non-document form of outputs as well.

5.2 Research Gap Based on Literature

The need for having a holistic view about the taxonomy and technology architecture in organizational transformation is highlighted in literature. It is also evident that taxonomy and technology architecture are key for any organizational change like *KM*. From the detailed literature survey, the research gap is shown in Fig. 5.1.

Development of taxonomy architecture for structured and unstructured knowledge for *KM* with navigation and content layer, development of metadata, and development of technology architecture with functional requirements, administrative/operational features, architectural features, integration features, user interface and navigation features, security features and documentation, and help features in an organization for the *KM* implementation are not widely explored in the literature. This taxonomy and technology architecture is very important for a manufacturing organization before the implementation of *KM* solution. Based on the development

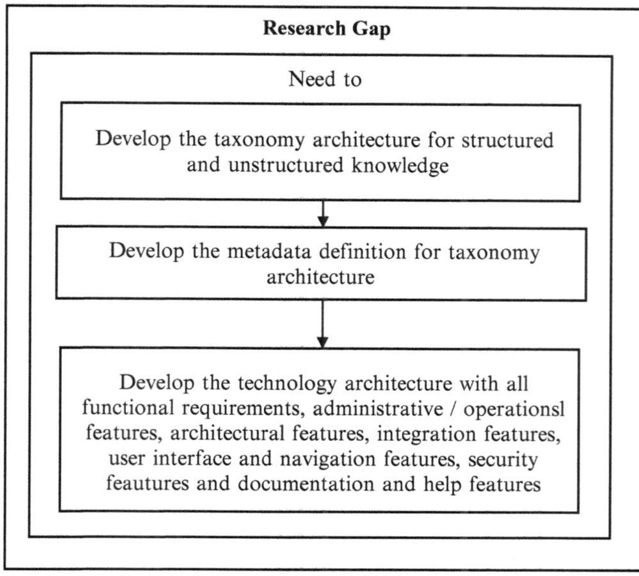

Fig. 5.1 Research gap

of taxonomy and technology architecture, organization should focus on process design and structure design for *KM*. Otherwise, implementation of *KM* solution will not be successful for any organization. The primary intention of taxonomy and technology architecture is to devise the backbone of *KM* which will be necessary for the implementation of *KM* portal. The objective of this module is to design a generic conceptual framework and generic design for taxonomy and technology architecture for any manufacturing organization. The factors related to taxonomy and technology architecture are derived from the literature, and those can be changed with respect to mission and vision of the organization.

5.3 Research Process and Methodology

The research process and methodology for the development of framework for taxonomy and technology architecture is detailed. The research process is divided into three phases:

Phase 1: Development of taxonomy architecture for structured and unstructured knowledge

Phase 2: Development of metadata definition for taxonomy architecture

Phase 3: Development of technology architecture with all functional requirements, administrative/operational features, architectural features, integration features, user interface and navigation features, security features and documentation, and help features

The research methodology used in the research process is detailed here:

In Phase 1, a Delphi-based detailed brainstorming exercise with a pre-intended taxonomy architecture which is derived based on business literature, research literature, individual discussions, and face validity with academic experts in the area of KM and consultants from the consulting organization in the domain of KM is the methodology. Delphi-based detailed brainstorming exercise is conducted with 43 executives belonging to 32 manufacturing organizations. For the derivation of pre-intended taxonomy architecture, discussions with 26 academic experts from 14 different top institutions all around India and 13 consultants from 4 different consulting organizations. These pre-intended process designs can be used as a base for any manufacturing organization. The devised taxonomy architecture is presented in Sect. 5.4 because the generic taxonomy architecture and the taxonomy architecture derived for the case study organization are the same.

The research methodology for Phase 2 and Phase 3 are similar to Phase 1. The pre-intended metadata definition and pre-intended technology architecture with all functional requirements, administrative/operational features, architectural features, integration features, user interface and navigation features, security features and documentation, and help features are devised based on the same methodology as indicated for Phase 1.

The devised metadata definition and pre-intended technology architecture with all functional requirements, administrative/operational features, architectural features, integration features, user interface and navigation features, security features and documentation, and help features are presented in Sect. 5.4 because the generic framework and designs and the framework and design derived for the case study organization are the same.

The diagrammatic representation of research design is detailed in Fig. 5.2.

5.4 Case Study Demonstration

The entire research process and methodology of taxonomy and technology architecture is demonstrated and applied through a real-life case study of Indian textile machinery manufacturing company. Thus, 216 executives of this textile machinery manufacturing organization were involved in a Delphi-based detailed brainstorming exercise for the development of all the designs of this study.

5.4.1 Taxonomy Architecture

The taxonomy architecture knowledge for case study organization was identified for both explicit/structured knowledge as well as tacit/unstructured knowledge.

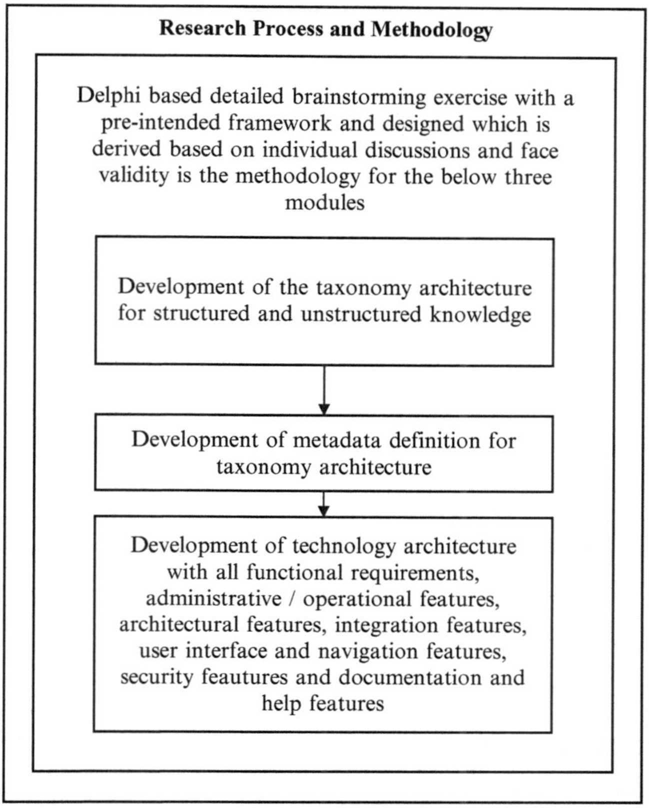

Fig. 5.2 Research process and methodology

5.4.1.1 Taxonomy Architecture for Structured Knowledge

For the structured knowledge, the taxonomy is classified into two layers, the navigation layer and the content layer. The navigation layer provides the access path to the information category as required by the user and the content layer facilitates a structured format for the storage and access of the right information. The navigation layer for taxonomy architecture of *KM* is devised and detailed in Fig. 5.3. The navigation layer with balanced scorecard perspective is devised and detailed in Fig. 5.4. The content layer for all the navigation layer taxonomy elements are devised and detailed in Figs. 5.5, 5.6, 5.7, 5.8, 5.9, 5.10, 5.11, 5.12, 5.13, 5.14, 5.15, 5.16, 5.17, 5.18, 5.19, 5.20, 5.21, 5.22, 5.23, and 5.24. The below sections show in detail the different layers of the taxonomy as applicable for case study organization.

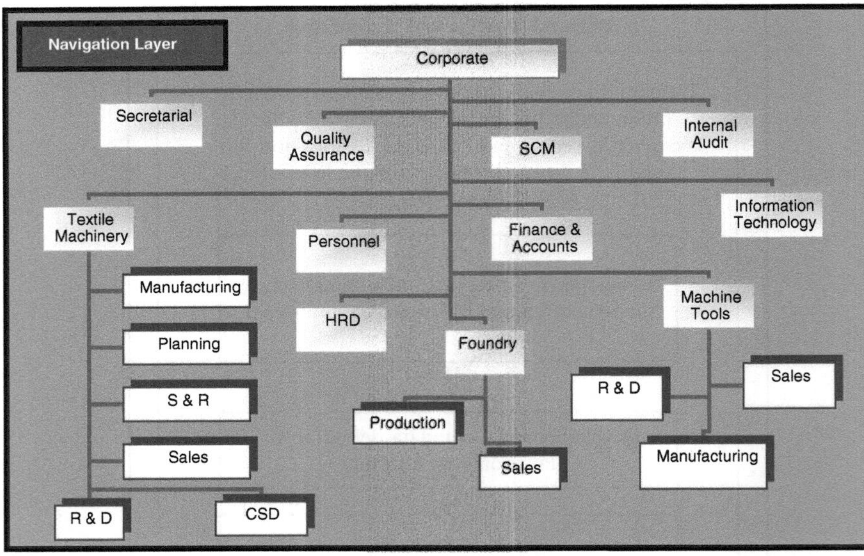

Fig. 5.3 Navigation layer for taxonomy architecture of *KM*

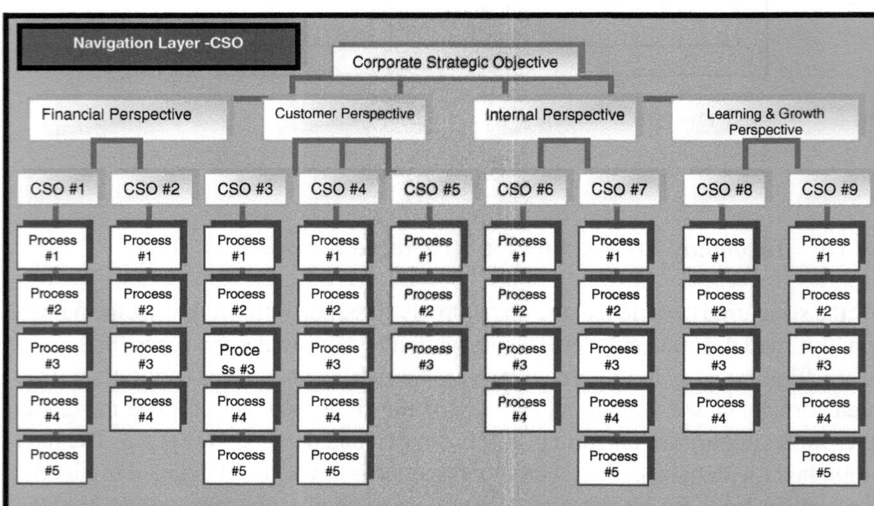

Fig. 5.4 Navigation layer with balanced scorecard perspective

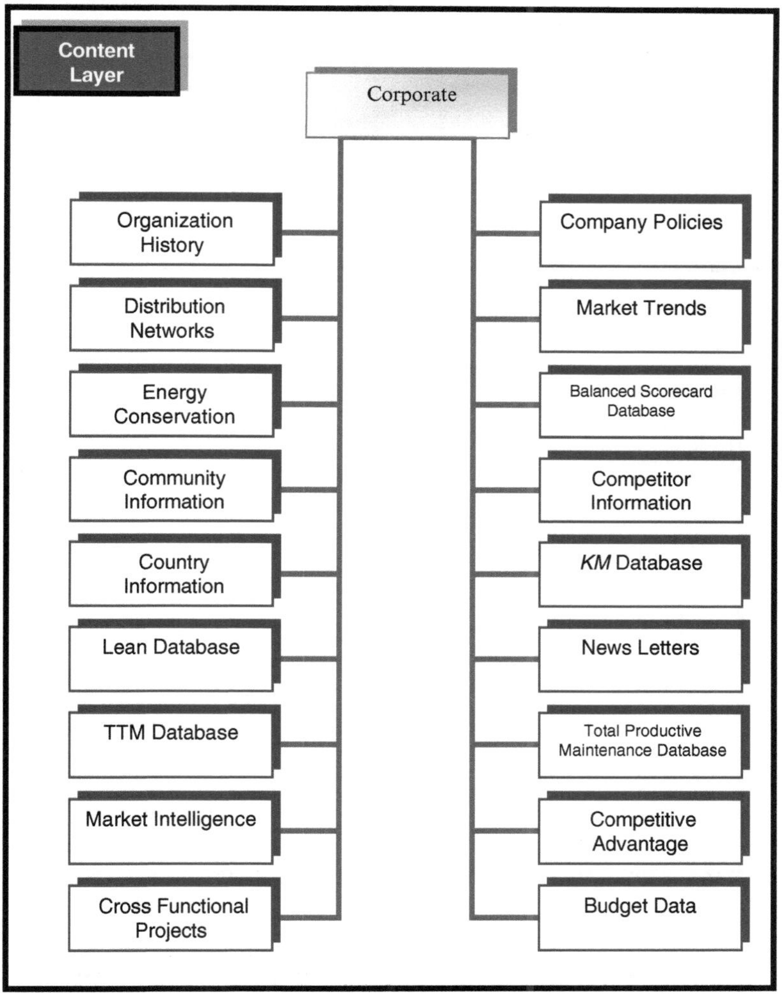

Fig. 5.5 Content layer for corporate taxonomy

5.4.1.2 Taxonomy Architecture for Unstructured Knowledge

Given that within case study organization there are several channels of interaction between employees that would normally contain tremendous amount of knowledge, it is important to extract and collect this knowledge in a codified form. The channels that need to be addressed currently are:

- E-mails/chat services
- Communities of practices/interest (organizational learning)
- Business and technical review meetings

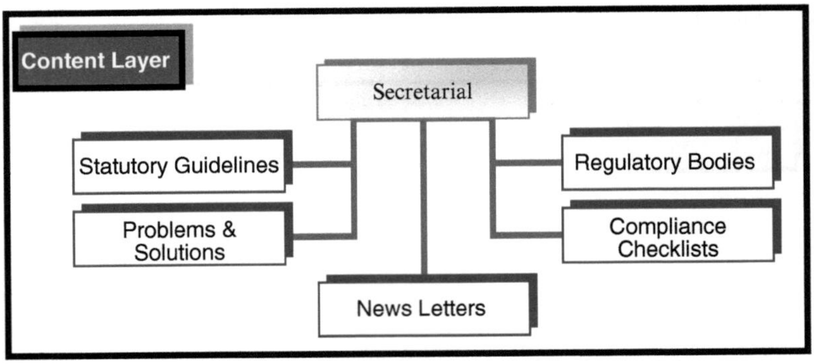

Fig. 5.6 Content layer for secretarial taxonomy

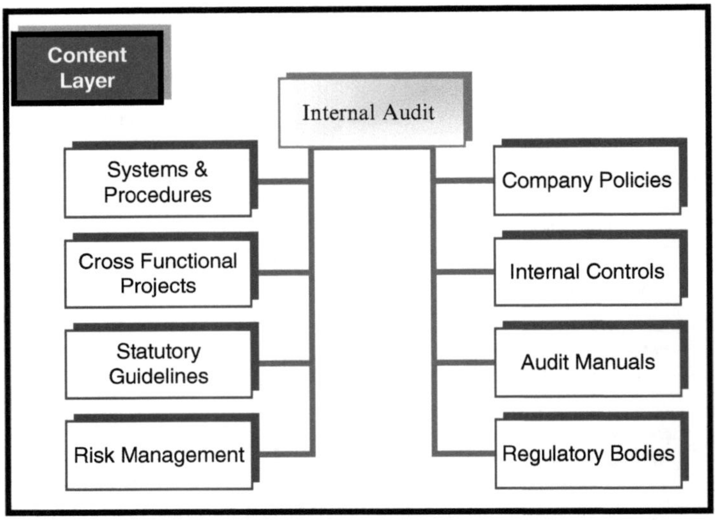

Fig. 5.7 Content layer for internal audit taxonomy

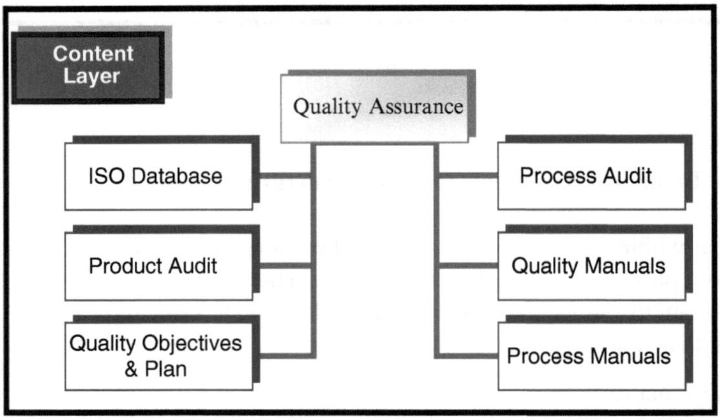

Fig. 5.8 Content layer for quality assurance taxonomy

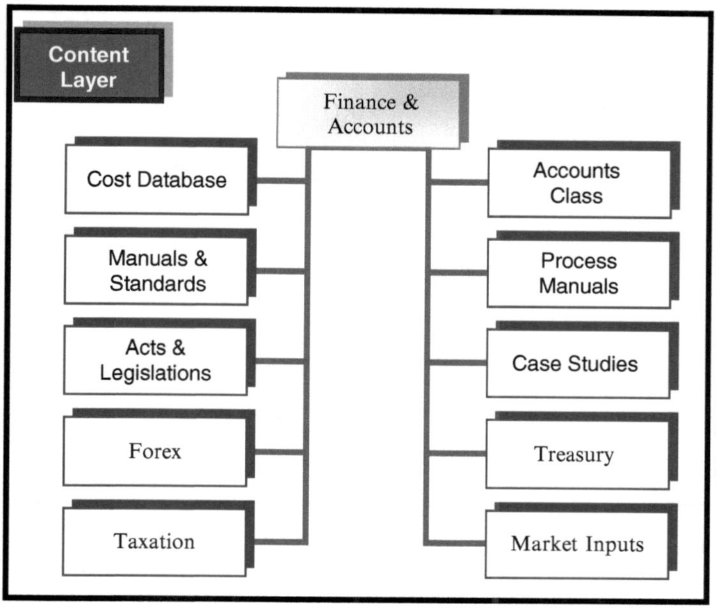

Fig. 5.9 Content layer for finance and accounts taxonomy

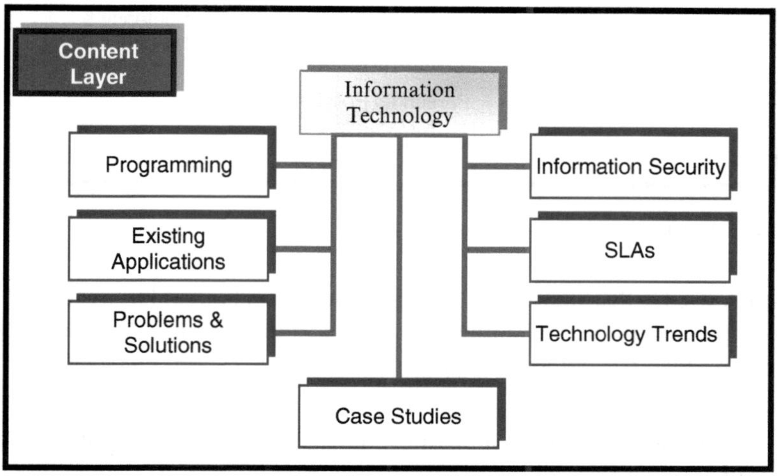

Fig. 5.10 Content layer for information technology taxonomy

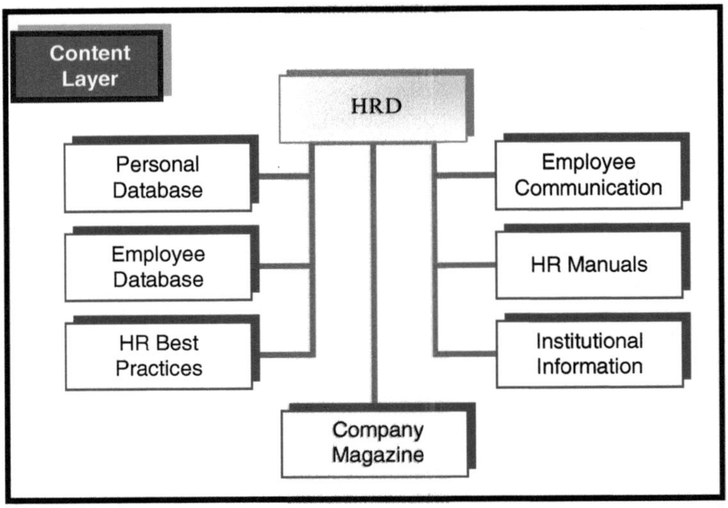

Fig. 5.11 Content layer for HRD taxonomy

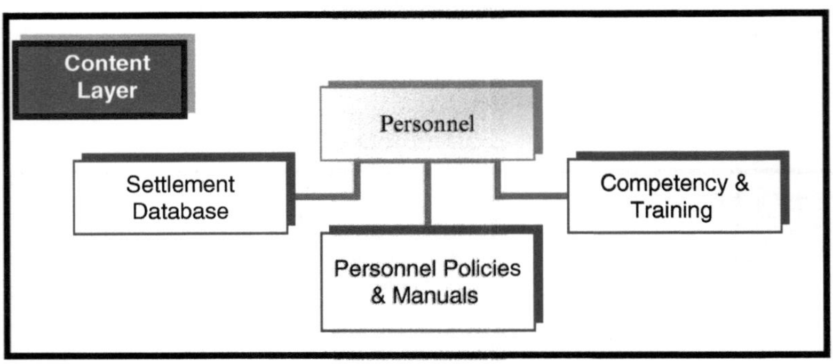

Fig. 5.12 Content layer for personnel taxonomy

The knowledge that we capture through these channels would be associated with at least one item in the structured taxonomy identified earlier. The unstructured knowledge would be codified to capture the following:

Queries: Answers and clarifications – This describes the queries/questions related to an item and the answers or clarification made available during these interactions on the channel:

1. Reference to other relevant structured knowledge components and industry benchmarks
2. Uniqueness of the query or situation to be highlighted

Symptoms: Root and cause (problems) – These are the set of observations discussed on a knowledge item feature with respect to its operational behavior, efficiency

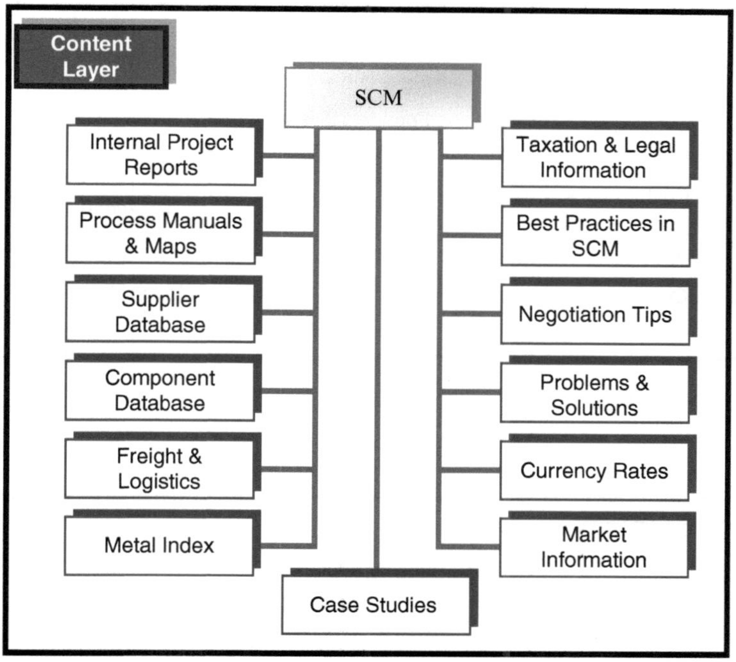

Fig. 5.13 Content layer for SCM taxonomy

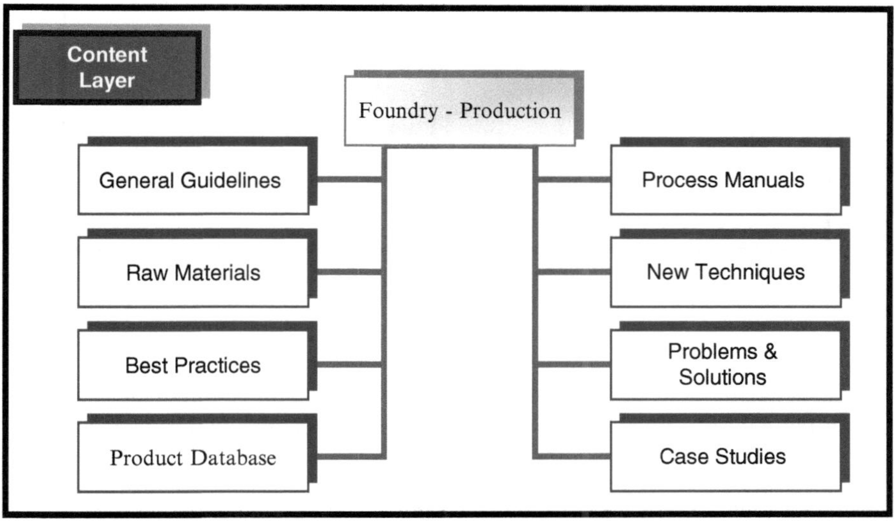

Fig. 5.14 Content layer for foundry-production taxonomy

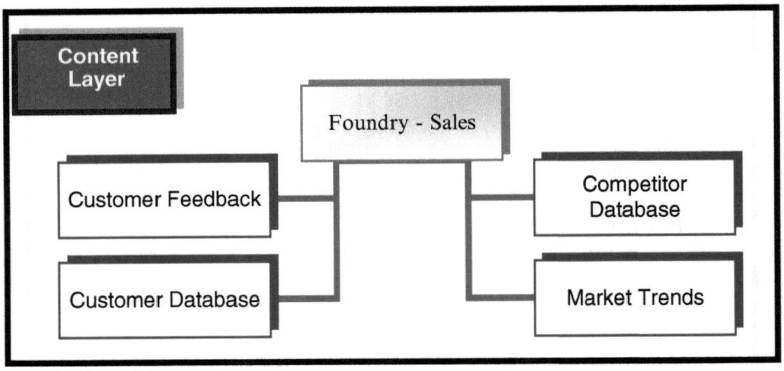

Fig. 5.15 Content layer for foundry-sales taxonomy

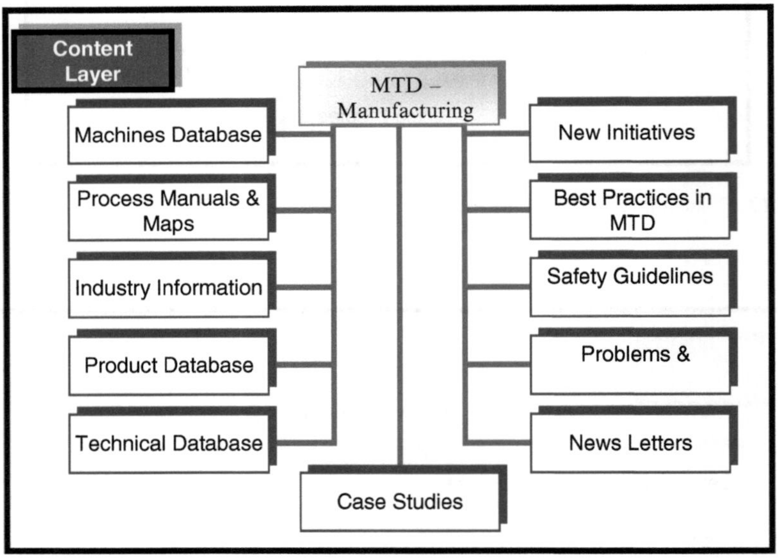

Fig. 5.16 Content layer for MTD-manufacturing taxonomy

characteristics, etc. In addition this would involve identification of the root cause for these observations, if discussed during the interactions in the channel:

- Information on the surrounding environmental factors, configuration parameters, and associated other happenings
- Domain-specific influence and related information

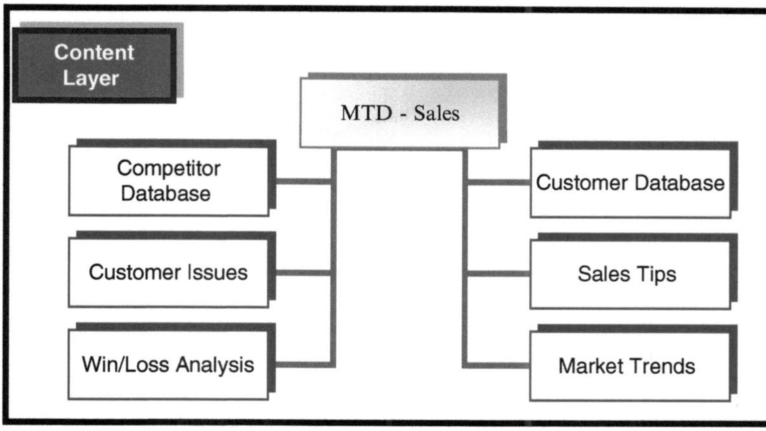

Fig. 5.17 Content layer for MTD-sales taxonomy

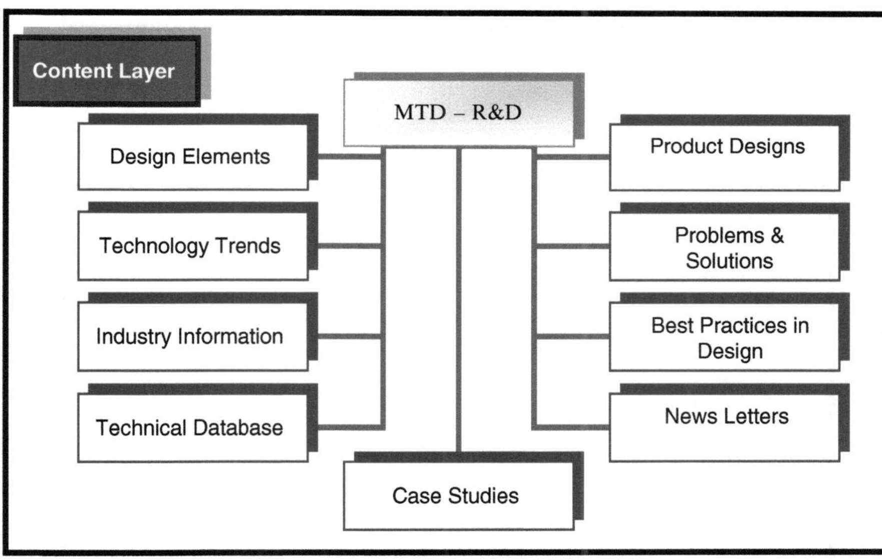

Fig. 5.18 Content layer for MTD-R&D taxonomy

Problems and resolutions – Essentially these are set of issues, concerns, or problem areas that get discussed on the channel for an item. Normally these are addressed to a designated organization expert or an informal area expert or a personal perceived expert. The response could normally be a resolution or a redirection to an expert or specific knowledge center:

(i) Alternatives in different situations
(ii) Associate in different functional areas or domains if applicable

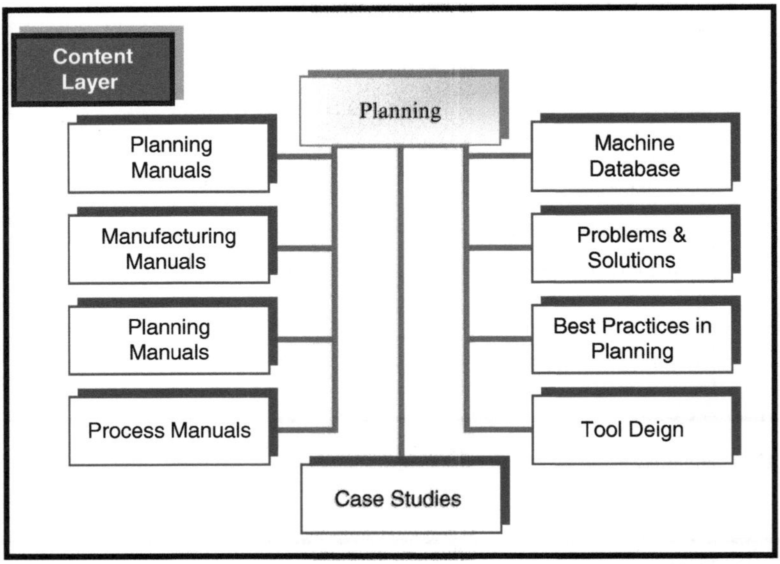

Fig. 5.19 Content layer for planning taxonomy

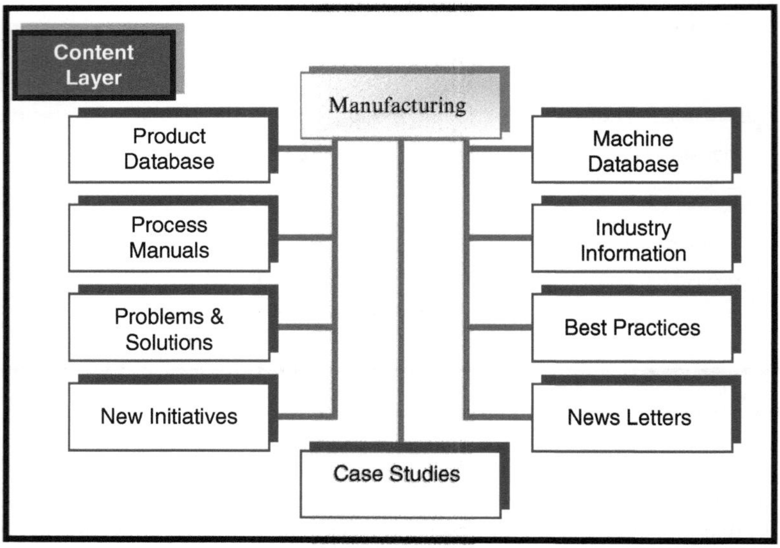

Fig. 5.20 Content layer for manufacturing taxonomy

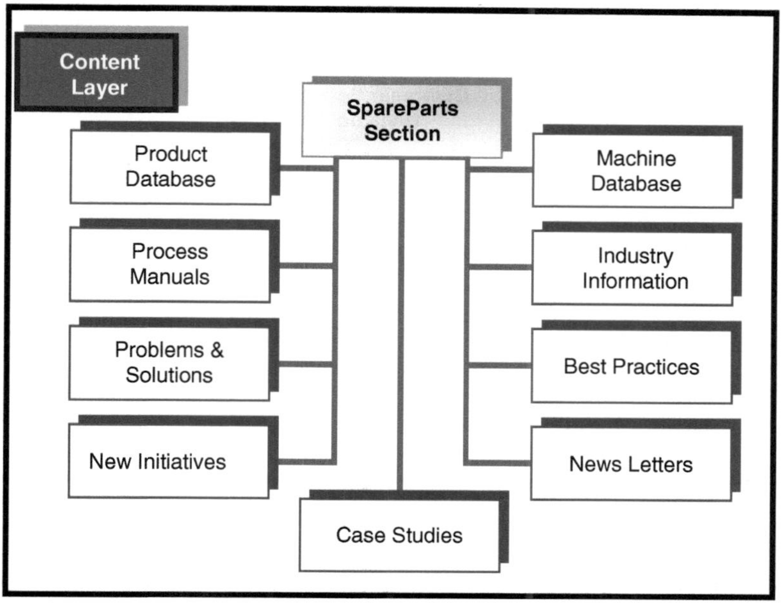

Fig. 5.21 Content layer for spare parts section taxonomy

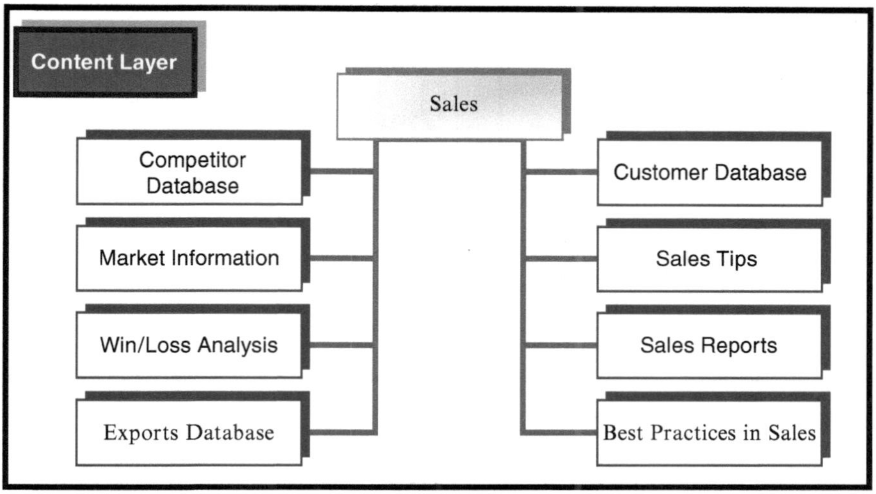

Fig. 5.22 Content layer for sales taxonomy

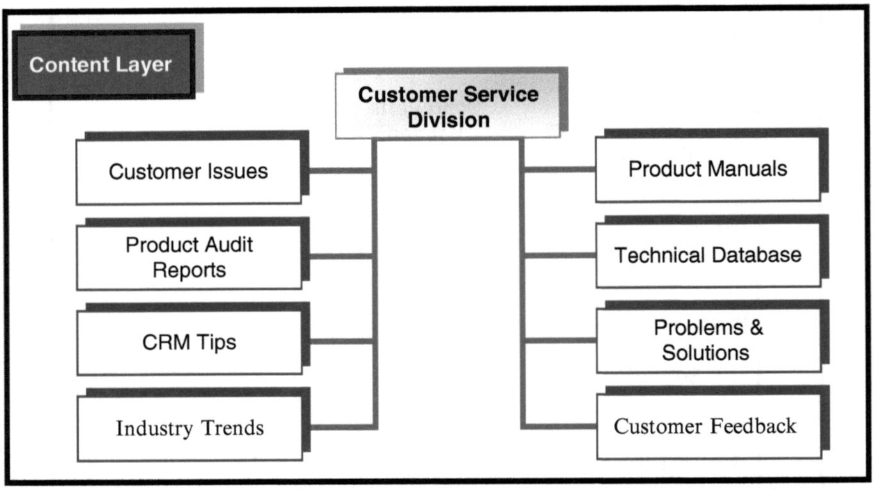

Fig. 5.23 Content layer for customer service division taxonomy

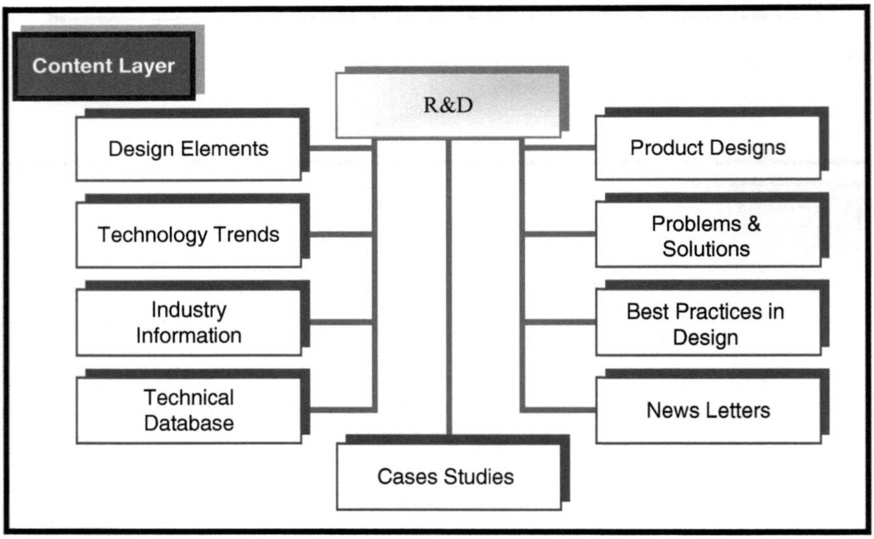

Fig. 5.24 Content layer for R&D taxonomy

Tips/suggestions and feedback – These are standard informal information floating around in the system as mails, chats, etc., as tips for the knowledge items in a functional area with some other knowledge worker substantiating or refuting the tip/suggestion with appropriate feedback:

1. Support with testified data or benchmarked industry source
   There are specific templates with respect to these unstructured knowledge items that can be used for any knowledge items across different service lines and vertical domains.

Table 5.1 Metadata definition of taxonomy architecture

S/No.	Attribute name	Attribute type	Remarks
1	Title	Text	Title of the document
2	Author	Text	Who wrote the document
3	Brief/abstract	Text	Brief description on the document
4	Categories	Multiple value	Category is the taxonomy navigation and basically will contain the elements as specified in the navigation layer of the taxonomy
5	Process	Text	From which process this knowledge item was created
6	Corporate strategic objective	Multiple value	To which corporate strategic objective (CSO) this document will contribute. This will be a drop down from the list of corporate strategic objectives as specified from the BSC
7	Subject	Multiple value	List of domains like textile machinery, machine tools, finance, and HR
8	Focus area	Multiple value	Related focus areas like business, technical, management, general, and training
9	Customer	Text	This property is optional. If the document is related with customer, we can specify the customer name
10	Understanding	Text	The document related to novice, expert, all, etc.
11	Rating	Numeric	The weightages for the document
12	Keywords	Multiple value	The search keywords for the document
13	Activities	Multiple value	To which activity this knowledge item is related
14	Form	Text	In what form the knowledge is captured doc, audio, video
15	Sensitivity	Single select	To what level the document is secured for access
16	Type	Text	What is the type of knowledge item ,e.g., report, best practice, and manual
17	Last revised	Date	Last modified

5.4.2 Metadata

Metadata are data that describe other data. Generally, a set of metadata describes a single set of data, called a resource. Metadata is structured, encoded data that describe characteristics of information-bearing entities to aid in the identification, discovery, assessment, and management of the described entities. Metadata is used to speed up and enrich searching for resources. Using metadata search with parameters such as the period, the category of knowledge, and author and title, standard query will be processed by taking context of the query into consideration. In general, search queries using metadata can save users from performing more complex filter operations manually. Table 5.1 defines the metadata that will be applicable across the taxonomy defined in the Sect. 5.4.1.

5.4.3 Technology Architecture

Technology architecture for case study organization is devised and detailed in Fig. 5.25. This section also discusses in detail the functional requirements and

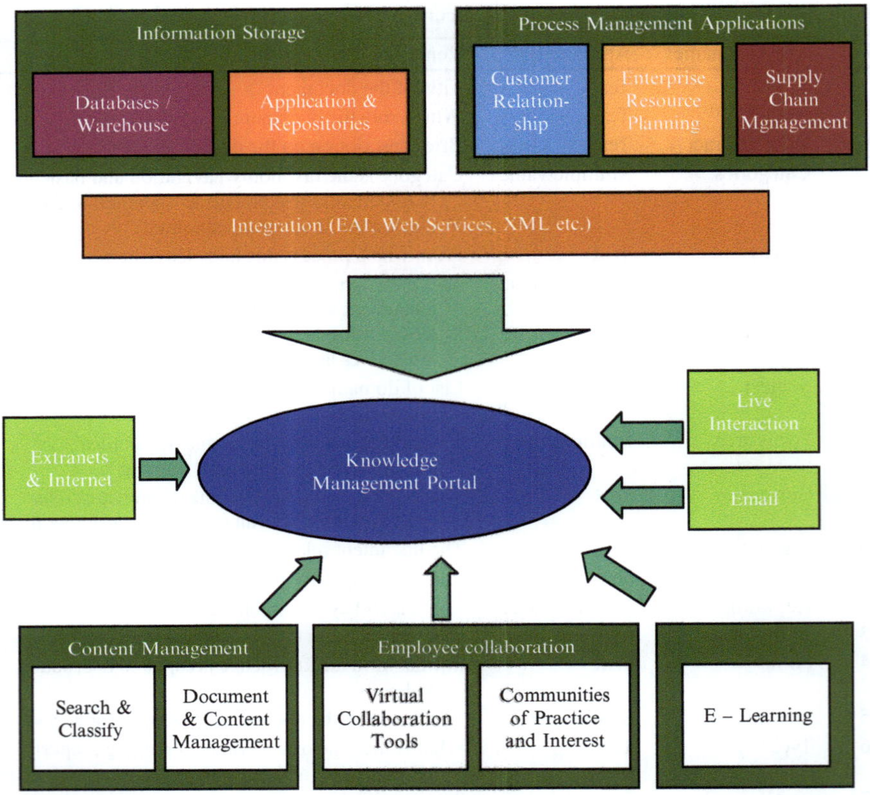

Fig. 5.25 Technology architecture for KM

features for the KM system. The requirements are classified under the following categories and the same are discussed in detail:

1. Functional features
2. Administrative/operational features
3. Architecture requirements
4. Integration requirements
5. User interface/navigation features
6. Security requirements
7. Documentation and help features

5.4.3.1 Functional Features

The functional features and description of functional features are detailed.

Taxonomy: Creation of multiple knowledge repositories and sub-repositories (typically, each repository would be an area of relevance to the businesses). Option for customization of taxonomy in order to dynamically adopt the changing requirements of the organization.

Clustering search: Quick, robust, and accurate search and retrieval capability. Users should be able to drill down into relevant categories. User to be responded with all the possible choices and ask to point which of the option he or she means. Wild and keyword-guided category and menu-driven search for distributed data/document or other sources.

Filtering – active filtering: Manually define filters and pointers to interesting content and share them across their group. Automated filtering: Statistical algorithms make recommendations based on correlation between user's personal preference and content rating. Content rating to be generated automatically by measuring the average time all readers spent on reading the item.

Contributor of the month: Display the image and name of the significant contributor to knowledge repository (automated statistical algorithms to make recommendations based on the parameters like number of contribution, quality, participation in unstructured channels).

User of the month: Display the image and name of significant user to knowledge repository (automated statistical algorithms to make recommendations based on the parameters like usage of information, participation in unstructured channels).

Integrative/structured knowledge: Shared medium for knowledge exchange for members of the same group to share, see, and contribute their knowledge. Collection of distributed knowledge repositories containing explicated (databases) and explicitly captured (talk/vision) content. Structured knowledge discovery facility to transfer knowledge from structured sources to people. (If the processes around it are well established, this can prove to be a valuable addition to the corporate sensory network as a way to learn about the business from the transactional interactions with customers, partners, and employees.)

Interactive/unstructured knowledge: Interaction among people and providing basic platform for people to share the tacit knowledge through various channels like:

1. Virtual meeting rooms
2. Discussion threads
3. Electronic white boards
4. E-mail integration (subject to security policies)
5. Videoconferencing

1. Virtual meeting room combines several forms of real-time communication, including voice over Internet protocol(IP), chat, and instant messaging. These forms are used to transfer knowledge from one person to another person or within small groups. Knowledge transfer using virtual meeting room is lost, much as it is with the telephone, because the participants never place a transcript or other record of their conversation in a repository.

2. Discussion thread: It can facilitate brainstorming and can act as a codification point for insights and lessons learned, and they can act as the tool through which distributed teams interact. Discussion threads must focus on the inquisitive dialog. In addition, discussion threads should be seen as a component of many of the other transfer technologies. Any real-time interaction, teaching session, or course should add a discussion thread that allows for asynchronous discussion to supplement the other interactions.
3. Electronic whiteboard: It supports the transfer of knowledge by allowing an instructor or peer to work out ideas on a shared white space. All members of the session may see the person work and in some cases may also collaborate, thereby transferring knowledge through peer/instructor interaction.
4. E-mail integration: It is a flexible, adaptive tool for the transfer of information and knowledge. E-mail is a private medium, however, so the learning that takes place between e-mail participants does so in a relatively private way. The only way knowledge is transferred on a larger scale is for the participants to decide to publish e-mail content or forward it to other people.
5. Videoconferencing: It has two modes. One, the speaker presentation mode, works well as a starting point for knowledge transfer in a lecture format, but if there are many participants, questions are better taken over the phone than through the video system because of the difficulties of managing windows and cameras. If the videoconferencing is also being used with video streaming, it is better for the audience and for the speaker to choose a single method of interaction, since some participants will not have cameras. The second mode is person-to-person or small-group-to-small-group interchanges where video can enhance the communication of knowledge by enhancing the overall bandwidth of the human interaction (facial expressions, body language, etc.). Videoconferencing in this mode is best when it is person-to-person and can be a very good supplement to mentoring if the mentor and student are not in the same location. It is also a good way to reduce travel when skill transfer is required because it not only allows people to communicate but it can also establish a relationship between the parties more quickly and with higher quality than phone, e-mail, or real-time chat.

Organization learning: The system should be able to showcase organizational learning in a structured format. Organizational learning is the process of "detection and correction of errors." Organizational learning occurs when groups of people give the same response to different stimuli:

1. Learning from experience
2. Learning from success/failure
3. Learning from external/new environment

Expert management: Mapping the expertise of an organization is valuable for several reasons. Easy access to a map of expertise of the organization can connect people when they need guidance resulting in quicker response rates and avoid rework. Maps can be used then to pull people in to assist on current projects or for offering training to employees who have existing good basic skills to equip them with additional skills the organization will need for future projects. Considerations include skills, expertise, experience, and location.

Some additional requirements in the same context are capability to provide links to "experts" in each repository/sub-repository and capability to post ideas/questions for experts.

Best practice sharing: The system should enable identification of best practices, documentation of best practices, and easy sharing of best practices.

News flash: Provide a quick idea of what is happening in the business environment internally and externally like business interest items, industry analysis, highlights, and lowlights.

What's new?: "What's new" button that displays new content added during, say, the last 2 weeks.

5.4.3.2 Administrative/Operational Features

The administrative/operational features and description of administrative/operational features are detailed.

Managing the *KM* site: The webmaster should be able to manage *KM* system and its repositories centrally. The administrator should be able to monitor the users and prevent hacking proactively and block the access for spurious/suspicious use. The administrator is to be given rights for the following:

- Setting rules
- Giving rights
- Modifying rights
- Removing access
- Uploading
- Content reliability
- Approval

Ownership/administration/content management of each repository by a different repository owner or knowledge champion to be enabled. Ability for all employees to upload (contribute) knowledge into repositories (which will be accepted after quality check by repository owners). Ability for all employees to download (reuse) knowledge from repositories.

Network management: The network should be managed from a central place.

Workflow capabilities: The system should have a step-by-step workflow capability in terms of document submission, alert to repository owner and editing, and acceptance for publishing in repository.

Reporting: System should be able to give logs/reports for no. of contributions, no. of hits per repository/sub-repository, no. of views/downloads per knowledge object, user names with names and dates of visit to the site, who has read/downloaded which knowledge object, who is the significant contributor for a period, and what are their contributions, report on obsolete contents on regular basis, etc.

5.4.3.3 Architecture Features

The architectural features and description of architectural features are detailed.

Performance: Any 300 KB to 1 MB files should not take more than 3 s to load under normal usage condition with maximum users online.

Scalability: The users are expected to grow. It should support an increasing number of users and higher load of transactions. Within acceptable limit for time delay for retrieval, query. Time for updates and inserts of new records should be minimum. Time delay for navigating between different parts of the interface to be minimum. Typically two third seconds is the maximum allowed.

Interoperability: Interoperability includes various factors such as:
Electronic mail: SMTP. X.400 and POP support
Documents: RTF, Microsoft office, PDF, etc.
Data access: SQL and ODBC
Internet: HTTP, XML, and FTP

Reliability: At any point transaction on *KM* server should give the correct/expected output. 99.99 % uptime. Time to rectify problem <30 min and standard audio and video supports.

5.4.3.4 Integration Features

The integration features and description of integration features are detailed.

With existing systems: Interface with existing Oracle Modules and Lotus Notes to get all structured knowledge in terms of extracting, structuring, storing, and viewing in different formats. Integration is primarily sought after with respect to the following: Technical Complaints Resolution History (Lotus Notes, "Metalink" – Oracle, Windows-2003), Service Request Threads (Oracle), e-mail services (Lotus Notes), and Messenger service ("Same-time" – Lotus Notes).

With new systems: Interface with new systems to be developed in the future to get all structured knowledge in terms of extracting, structuring, storing, and viewing in different formats.

5.4.3.5 User Interface and Navigation Features

The user interface and navigation features and description of user interface and navigation features are detailed.

Functionality: The system should be user-friendly and users should be able to accomplish their task quickly, effectively, and without frustration over the systems usability. The system should take care of end-user needs and requirements.

Consistency: There should be consistency across all parts of the *KM* system in the way in which information is presented, accessed, and used like MS office.

Visual clarity and layout: The system should enable users to easily find the information they need. Present all information that relate to the user's task on one screen if possible and hide all unrelated information and control by default. Have sufficient white space on the screen and use lowercase for dense area. Use the right font and font size. No cluttering on the screen. Hyperlinks to provide further information on text string.

Navigation and control: Site map to be provided. The user should be able to tell which area/tools he or she is using at a given moment.

Images: Interesting and relevant images to be used at appropriate places.

Feedback: User should be able to receive the feedback from the system so that they know what the system is doing and what is expected next. Audible clues/alerts to be provided.

Personalization: User should be able to configure his or her user ID according to one's knowledge requirement. Individual users may subscribe to selected repositories that are of relevance/interest to them; alerts through e-mail when there is a new addition to the selected repositories to be enabled.

Rating: The readers should be able to rate an article/paper/resource.

Knowledge points: Knowledge currency/points, etc., for knowledge submission each time somebody opens/downloads (other than the person who has submitted).

5.4.3.6 Security Features

The security features and description of security features are detailed.

Access control: File level access control with user groups. User groups are classified into U1, U2, and U3 for businesses and functions. U1 will have access right to all the files. U2 will not have access to certain sensitive files. U3 will have access only to limited files. *KM* administrator will assign the user access level for each file while uploading so that the user can have access to all the files that they are authorized to view. Change of user access level should be possible at any time.

External access: Capability to provide access to stakeholders other than employees (e.g., distribution agents, customers, partners, suppliers) to selected parts of the *KM* portal. Capability to provide links to external content/sites, etc.

5.4.3.7 Documentation and Help Features

The documentation and help features and description of documentation and help features are detailed.

Documentation: Complete design, development, and implementation should be documented at every stage. All test results should be documented. Modifications/changes should be approved and documented.

Help: The system should have a self-help for various features and how to use it with tips, cautions, etc.

5.5 Summary

The generic taxonomy and technology architecture developed in this research can be directly taken as base for any manufacturing industry in building *KM* solution, and the practice managers may concentrate on the various components. As the foundation for all activities within the corporation relating to explicit and tacit knowledge, a taxonomy can further a wide range of corporate objectives, such as enabling business processes, protecting intellectual property, and building the foundation for compliance. Each organization requires a different taxonomy because each has unique processes, organizational configurations, core competencies, and histories.Reference

Chapter 6
Process Design for Knowledge Management Solution Implementation

6.1 Introduction

Most of the multinational companies aim to embark an initiative to enhance the capture, storage, and dissemination of knowledge and information. The main objectives behind this initiative are the identification and analysis of available and required knowledge assets and knowledge asset–related processes and the subsequent planning and control of related actions to develop both the assets and the processes so as to fulfill organizational objectives.

In an organizational context, data represents facts or values of results, and relations between data and other relations have the capacity to represent information. Patterns of relations of data and information and other patterns have the capacity to represent knowledge. For the representation to be of any utility, it must be understood, and when understood the representation is information or knowledge to the one that understands.

KM is the process through which organizations generate value from their intellectual and knowledge-based assets. Most often, generating value from such assets involves sharing them among employees, departments, and even with other companies in an effort to devise best practices. KM is often facilitated by information technology (IT), though technology by itself is not KM.

Knowledge assets are the knowledge regarding markets, products, technologies, and organizations that a business owns or needs to own and which enable its business processes to generate profits, add value, etc. KM is not only about managing these knowledge assets but managing the processes that act upon the assets. These processes include developing knowledge, preserving knowledge, using knowledge, and sharing knowledge.

KM can be effectively implemented by proper knowledge organization structure and carried forward by incentive plan and lucrative reward programming. To better implement KM, some organizations place the positions entitled with, for instance, chief knowledge officer (CKO), knowledge engineer, knowledge analyst,

K. Ganesh et al., *Design and Development of Knowledge Management for Manufacturing: Framework, Solution and Strategy*, DOI 10.1007/978-3-319-02892-7_6,

knowledge manager, and knowledge steward to administrate KM. And along with the development of information technology (IT), KMS has been integrated in organizational structure to assist in managing knowledge through intranet or Internet.

6.2 Research Gap Based on Literature

The need for having a holistic view about the process design, reward design, and communication plan in organizational transformation is highlighted in literature. It is also clearly evident that all the three designs are the key for any organizational change like KM. From the detailed literature survey, the research gap is shown in Fig. 6.1.

Development of process design for the knowledge capture, knowledge storage, knowledge retrieval, and knowledge administration; development of key performance indicator framework; development of organizational structure for KM with roles and responsibilities; development of reward design with incentive plan and metrics; and development of communication plan design in an organization for the KM implementation are not widely explored in the literature. This process design with reward and communication plan is critically important in an organization before the implementation of KM solution. Based on the development of process

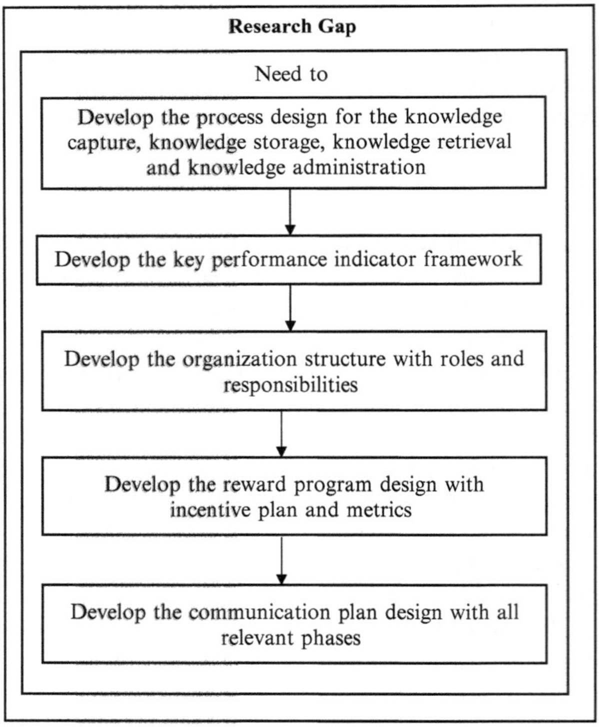

Fig. 6.1 Research gap

design, reward, and communication plan, organization should focus on learning design and development of linkage with internal and external functions of organization. Otherwise, implementation of KM solution will not be successful for any organization. The primary intention of process design, reward, and communication plan is to devise the backbone of KM which will be necessary for the implementation of KM portal. The objective of this module is to design a generic conceptual framework and generic design for process design, reward, and communication plan for any manufacturing organization. The factors related to process design are derived from the literature and those can be changed with respect to the mission and vision of the organization.

6.3 Research Process and Methodology

The research process and methodology for the development of framework for process design, reward plan, and communication plan is detailed. The research process is divided into five phases:

Phase 1: Development of process design for the knowledge capture, knowledge storage, knowledge retrieval, and knowledge administration
Phase 2: Development of key performance indicator framework
Phase 3: Development of the organization structure with roles and responsibilities
Phase 4: Development of reward program design with incentive plan and metrics
Phase 5: Development of communication plan design with all relevant phases

The research methodology used in the research process is detailed here:

In Phase 1, a Delphi-based detailed brainstorming exercise with a pre-intended process design which is derived based on business literature, research literature, individual discussions, and face validity with academic experts in the area of KM and consultants from the consulting organization in the domain of KM is the methodology. Delphi-based detailed brainstorming exercise is conducted with 43 executives belonging to 32 manufacturing organizations. For the derivation of pre-intended process design, discussions were conducted with 26 academic experts from 14 different top institutions all around India and 13 consultants from 4 different consulting organizations. This pre-intended process design can be used as a base for any manufacturing organization. The devised process design is presented in Sect. 6.4 because the generic process design and the process design derived for the case study organization are the same.

The research methodology for Phases 2, 3, 4, and 5 is similar to that of Phase 1. The pre-intended key performance indicator framework, pre-intended organizational structure with roles and responsibilities, pre-intended reward design with incentive plan and metrics, and pre-intended communication plan design are devised based on the same methodology as indicated for Phase 1.

The devised key performance indicator framework, organizational structure with roles and responsibilities, reward design with incentive plan and metrics, and

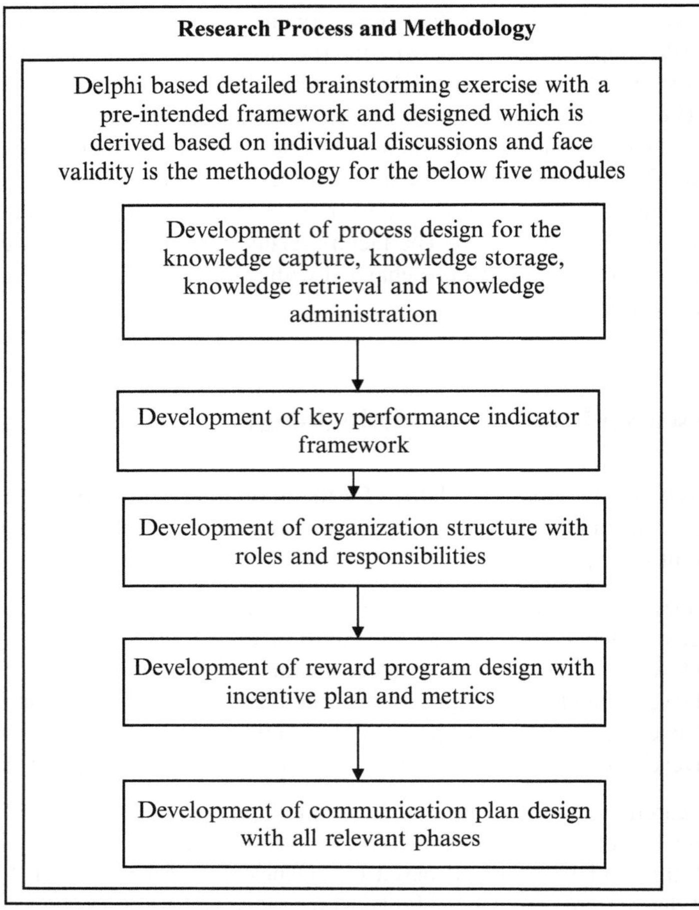

Fig. 6.2 Research process and methodology

communication plan design are presented in Sect. 6.4 because the generic framework and designs and the framework and design derived for the case study organization are the same.

The diagrammatic representation of research design is detailed in Fig. 6.2.

6.4 Case Study Demonstration

The entire research process and methodology is demonstrated and applied through a real-life case study for Indian textile machinery manufacturing company. Thus, 216 executives of this textile machinery manufacturing organization were involved in Delphi-based detailed brainstorming exercise for the development of all the designs of this study.

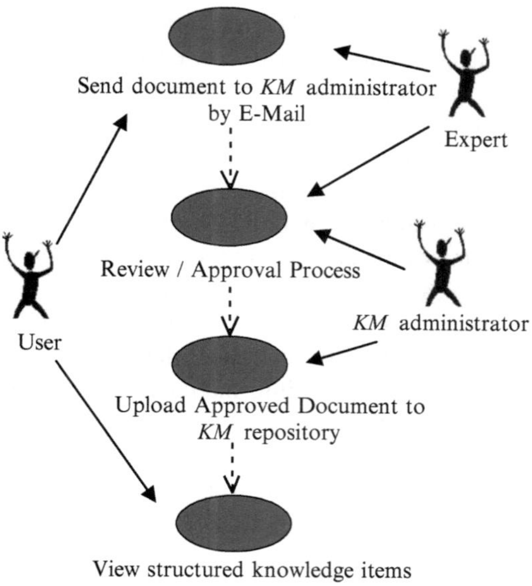

Fig. 6.3 Process environment design – structured knowledge

6.4.1 *Process Design Development*

This section will define the KM processes for the case study organization that needs to initiate in order to achieve its KM goals and objectives. The processes thus described revolve around the following four broad areas which form the main process for any KM initiatives such as knowledge capture, knowledge storage, knowledge retrieval, and knowledge administration.

6.4.1.1 Knowledge Capture

One of the initial steps toward KM is to capture the knowledge that exists in the organization. The different processes for knowledge capture that can be adopted in a manufacturing organization are defined with respect to the following three areas, namely, structured knowledge, document workflow, and unstructured knowledge.

Structured knowledge: This section defines the process to capture and store all the structured information (knowledge items) in a manner such that these can be easily retrieved at any point in time. The process environment design for structured knowledge is explained in Fig. 6.3.

Any KM tool provides the facility to upload content and catalogue it using attributes. These attributes are used when doing selective searches. There are a number of ways in which content can be collected for storing into the KM repository. The following procedure is suggested for capturing and storing structured knowledge. These are:

- A user voluntarily submits a document to KM through an offline system such as e-mail for upload into KM. In this case KM administrator would review the document and upload it after assigning the required attributes, using the content management tools' content upload workflow. The contribute section captures the categories, to which a knowledge item is uploaded and also the attributes related to that category from the user who uploads the document to the system.

- An end user can upload content directly into KM. Here (based on the licensing issues), the user would upload the content into the content upload workflow of the document management tool, and the content would get uploaded after it is approved by the expert/KM administrator.
- Some of the contents in the KM repository can be collected directly from other applications. Such data can be made available (for search and access based on security policies) in KM that are residing in internal application databases or external links. In such cases the KM system should be able to talk to these databases and retrieve information, on an "as and when" required basis.
- The workflow process will facilitate KM administrator in ensuring that the knowledge items are available as and when they are due. The detailed explanation of the document workflow process is mentioned in the "document workflow" section.

Key roles involve KM administrator, KM expert, and user.

Document workflow: This workflow is required to push the knowledge items contributed by users through a review and approval mechanism before they get published in KM site. The knowledge items contributed to the KM through contribute section are routed through a workflow process. The participants in the workflow are expert and KM administrator.

Expert: The contributed documents to KM site will appear in the expert section for approval. The expert for each category is identified and should be able to see the documents concerning his area. The expert should be able to do the following actions such as approve, reject with reason, change profile values, and change categories.

Administrator: After the knowledge item successfully passes the expert approval process, it will be waiting for KM administrator to publish it for display in KM site as shown in Fig. 6.4. Administrator should be able to do the following actions such as publish, reject with reason, change profile values, and change categories.

Key roles involve KM administrator and KM expert. The process environment for administrator is explained in Fig. 6.4.

Unstructured knowledge: This section defines the process to capture and store the unstructured knowledge that is being exchanged through unstructured means such as chats and discussion forums in a manner such that these can be easily retrieved at any point in time and are also made available for searches across the KM site.

Employee chat: This is an online chat facility that is required. The chats can be mainly many-to-many chat or one-to-many chat (chat with an expert) or one-to-one chat as in Fig. 6.5. In all cases, there should be an authentication of the person entering into the chat. All chats should be moderated. Any user can request for a chat

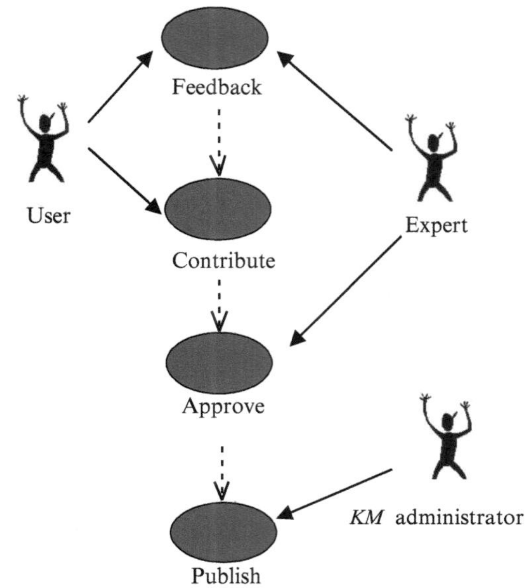

Fig. 6.4 Process environment design – administrator

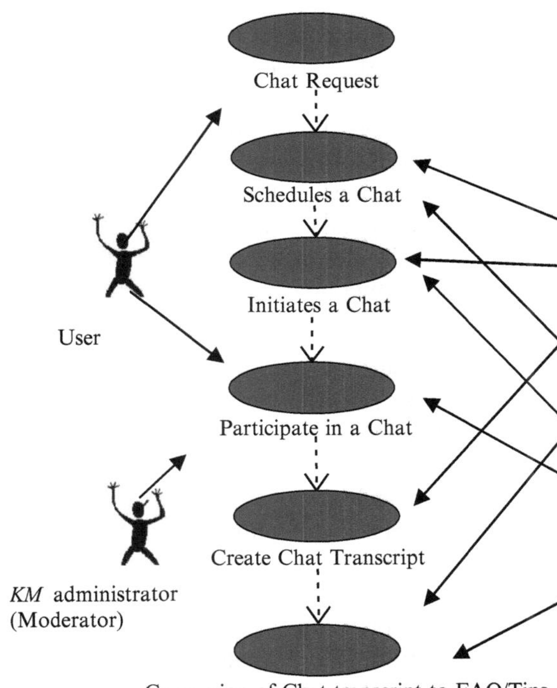

Fig. 6.5 Process environment design – employee chat

Fig. 6.6 Process environment design – discussion forums

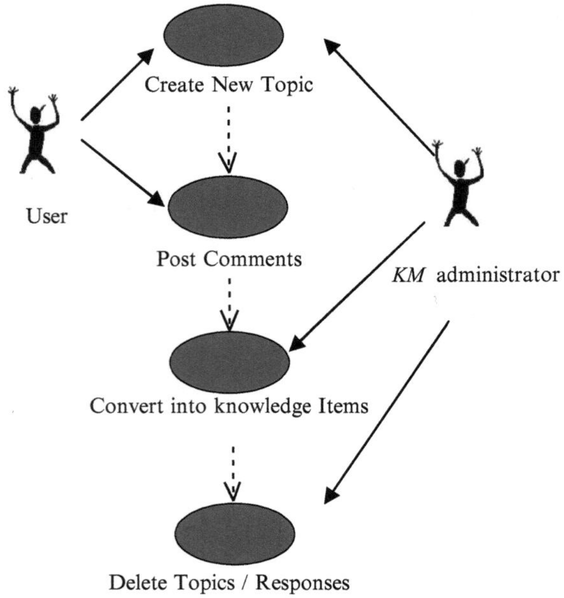

along with the information on the topic and time for the chat. The KM administrator schedules the chat with respect to the request from the user or on his own and invites users for the chat. He is the owner of the chat. The owner initiates the chats at the scheduled time and intimates the invited users. There should be a facility of creating a knowledge item out of a chat transcript. This knowledge item is stored in the KM repository with the appropriate properties tagged to it and appropriate security settings. The KM administrator can also create frequently asked question (FAQ)/tips out of chat transcript. The expert should be able to select questions to edit and submit to the new document, FAQ or tips. The process environment design for employee chat is explained in Fig. 6.5.

Discussion forums: This is essentially a bulletin board service. The KM administrator should be able to create forums under particular domains and sub-domains. He can modify users list at any point of time. Under such forums the administrator can create a new topic and can also approve/reject topic requests from users for discussion. Any assigned users can then post their views on this topic. The thread will continue with users' replies against topics and also with replies to the replies. Only KM administrator will have the facility to delete topics or the replies posted as explained in Fig. 6.6. View access will be available for all assigned users to the forums, whether they have posted a question/comment/reply or not. There should be a provision for the topic to expire at a given time. KM administrator will then convert it (offline and manually) into useful knowledge items, to be stored in the repository, if required, else the entire transcript can be stored in archives. The process environment design for discussion forums is explained in Fig. 6.6.

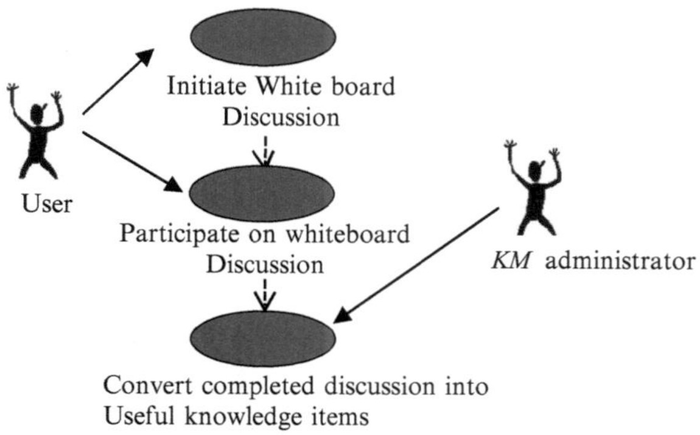

Fig. 6.7 Process environment design – online collaboration

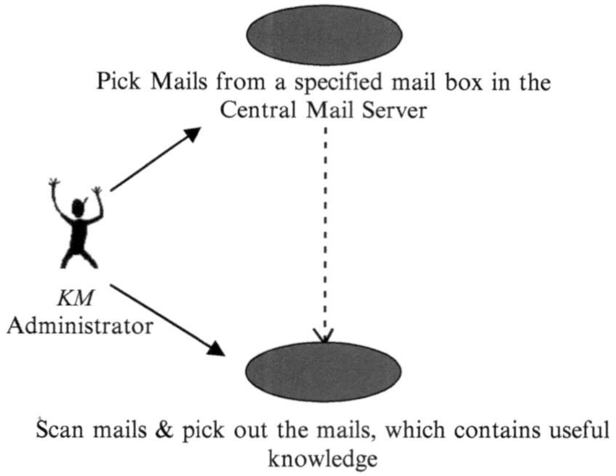

Fig. 6.8 Process environment design – e-mail integration

Online collaboration: In this process, there will be a common "whiteboard" on which different users who have logged in can share views online as shown in Fig. 6.7. The whiteboard will allow users to draw (like a paint brush in MS Office) on this common screen and that image will be seen by all logged in users. This way a cumulative effort can be used to develop a concept/strategy, etc. The system will store the completed discussion and will be available with KM administrator for further use. The process environment design for online collaboration is explained in Fig. 6.7.

E-mail integration: The system is the same as that described in structured knowledge. In addition, the system should provide a facility to be able to pick mails from the central mail server as explained in Fig. 6.8. All mails addressed to a designated mail box for KM would be stored as a copy in the mail server. The KM system should be able to scan through these mails in the specified mailbox and search for any key "knowledge items" and bring out the search-specific e-mail contents. The key role include KM administrator, KM expert, and KM user. The process environment design for e-mail integration is explained in Fig. 6.8.

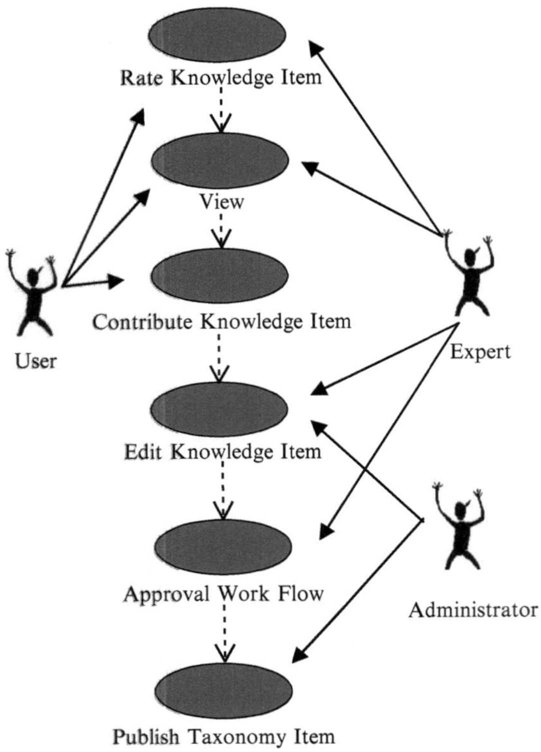

Fig. 6.9 Process environment design – knowledge storage

6.4.1.2 Knowledge Storage

This section focuses on the storage of knowledge items as per the taxonomy and metadata definitions. Taxonomy defines the way the data is stored in the back end systems of the KM repository. It gives a classification of the way the data is catalogued as shown in Fig. 6.9. The knowledge items present in KM site should be linked to more than one category logically. This is as defined in the metadata definitions. Access control to the document should be established to the user level. The primary requirement here is to provide for distributed data content with authorization and access control. The key roles include KM administrator and KM user. The process environment design for knowledge storage is explained in Fig. 6.9.

6.4.1.3 Knowledge Retrieval

This is the process which explains how a user should be able to search and filter the contents on a variety of parameters, so as to finally reach the knowledge items desired by him. The results will be based on taxonomy and metadata definitions.

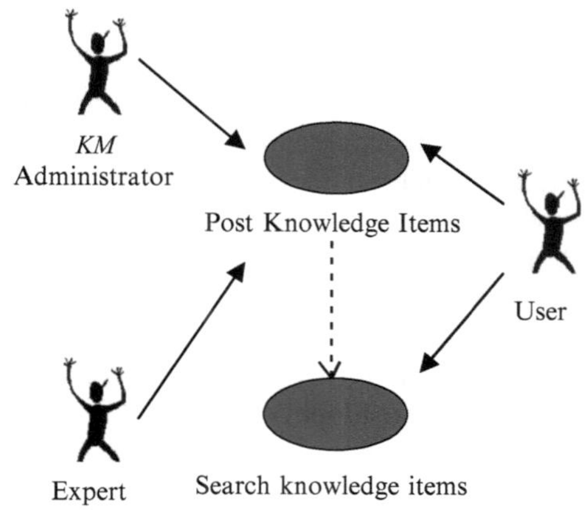

Fig. 6.10 Process environment design – expert page

The concept of search and filtering is producing a relatively small subset containing a high proportion of relevant documents, whereby the user can locate the required information quickly. The search and filtering could be on the following parameters such as keyword, taxonomy, search in document indices, and expert search. Keyword is the basis of the search. In the search, the entered keywords would be matched to the starting strings of the article keywords. Taxonomy is a search that could be done only on selected domain (taxonomy section). The list of domains will be maintained by the KM administrator as a master list. The users will get a pick list to select one or more domains. Search in document indices involves the catalogued document attributes or metadata along with the full-body text search. If "metadata" is selected, then the entered keywords would be searched in the "title, abstract, keywords" fields of all the documents in the KM database. These fields will have to be entered when content is being uploaded into the system. In case of a "full-text search" the system would search through the contents of the documents and articles subject to the suitable search engine being used. After the user submits the selection, the result page would be displayed. The resulting articles would be categorized based on the prevailing taxonomy and displayed in descending order of the last modified date in each subsection. In the first page after the search, the number of results found in each section would be listed. The title of the section would be hyperlinked, and on clicking on the title, the full list of knowledge items in that section would be listed in a new window. The user can then view/download the individual knowledge items based on his access rights. Expert page is based on the keywords entered by the user for the search; the system will also search the catalogued expert pages for the matching keywords. If identified, links to those expert pages would also be shown on the search results page. To allow this link to be established, the KM administrator would need to assign the keywords for each expert in the expert definition databases as shown in Fig. 6.10. The key roles include KM administrator, KM expert, and KM user. The process environment design for expert page is explained in Fig. 6.10.

6.4.1.4 Knowledge Administration

This section defines the activities/functions that the KM administrator team has to perform to maintain KM. The activities are defined section-wise below. The current taxonomy definitions are based on the discussions with business executives of manufacturing organization, and it is only representative of the present business scenario. It is quite likely that the taxonomy will undergo changes in the course of time as the KM gets refined with use in manufacturing organization. In case there is a change in the taxonomy, KM administrator would need to move the contents accordingly. KM administrator could do the following tasks such as add new taxonomy items, delete taxonomy items, and move contents from one head to another head. Retrieval is nothing but the KM administrator will have to continuously track the usage of the contents in KM and accordingly create or modify the domains/categories in the master list. This will have to be done primarily with the view of making it easier for the users to access the right information quicker. It is the task of the KM administrator to facilitate the incentive plans for KM. KM administrator with the help of the administrator module should do the following tasks such as generate reports on the contributions and usage for the appropriate month/quarter and level of the organization and run the report, select the contributor/user of the month, and plan and facilitate to provide the incentives for the respective outcomes. The KM administrator will receive inputs from the users/experts by e-mail or hard copies of structured knowledge items (office files or PDF files) on a regular basis. The KM administrator would need to do the following tasks such as upload the content with set of attributes; delete the content from the repository when required; convert chat and bulletin board transcripts into FAQs, best practices, problems, and solutions and upload into relevant portions of the repository; decide the relevant e-mails collected and convert to "knowledge" and store in the repository; and follow up for the knowledge items as part of the process tracking and upload into the repository. The output from this process is the uploaded content that will appear in the KM repository under selected taxonomy item. The inputs for unstructured knowledge process are the chats, discussion forums, etc., which are defined in the unstructured knowledge capture process. The following tasks would be performed by the KM administrator as part of this process:

- Employee chats:
  - Set up chat sessions with expert users and inform to all concerned.
  - Invite concerned persons for the chat.
  - Moderate chat sessions.
  - Review a chat transcript and convert to a FAQ and post into the repository.
  - Follow up for answering of the open questions in the chats.
- Discussion forums:
  - Create a new topic for discussion.
  - Delete expired topics and irrelevant responses (at least for topics owned by KM).

  - Review the whiteboard discussions and post into KM repository.
  - Review the e-mails for knowledge items and post into the KM repository.
- Convert online collaboration discussions into FAQs and post in the KM repository.
- Housekeeping activities to delete old/irrelevant content.

This process will not only facilitate the capture and storage of unstructured knowledge within the manufacturing organization but also ensure that relevant information reaches the concerned employees about chats and discussions in the form of mails and flashes on the site so that employees are kept aware of the happenings.

Organization learning process provides for identifying the knowledge items related to organization learning and posting the same into the appropriate sections. The primary objective of this process must be to upload relevant content and tag it to be visible on the organizational learning section of the KM site. As part of expert management process, the KM administrator would need to do the following tasks:

- Expert identification
- Uploading of all expert's detail to the site
- Mapping of subjects/topics to expert
- Conversion of expert chats to FAQ
- Feedback rating and review to maintain expert's rating
- Additional inputs to expert to make the expert pages complete

The primary objective of this process must be to identify and upload relevant information on experts in the organization on various subjects and tag it to be visible on the expert management section of the KM site.

6.4.2 Key Performance Indicators (KPI) Design

As part of this process, the KM administrator would be responsible to track the performance of the KM process on a regular basis. The administrator will use the reporting features of the KM system to generate reports specific to some key performance indicators (KPI) as defined below. These reports will then be analyzed and projected to the management as inputs for decision making and further improvements/changes to KM.

The KPIs that need to be tracked for KM are:

1. System-specific KPIs:

   (a) Number of log-ins into the KM portal
   (b) Number of hits per section/subsection
   (c) Number of views/downloads per knowledge item
   (d) Number of contributions

2. Corporate strategic objective (CSO)-related KPIs:
   (a) Number of documents related to each of the CSOs
   (b) Exception reports for tracking CSO which do not have any knowledge items or have consistently low (less than 5 per month) knowledge inputs
3. Expert performance indicators:
   (a) Number of reviews/approvals by the expert
   (b) Number of pending reviews/approvals
   (c) Mean time taken by the expert to review and approve a document
4. Significant contributors/users for a period and what are their contributions. In this context significant contributors/users would be the first 15 top-rated contributors/users.
5. User names with names and dates of visit to the site, who has read/downloaded which knowledge item.

The management should set targets against each of the above KPIs. The following guidelines can be used to initiate the measures initially. Before going live ensure that there is a balance of knowledge items available in the system across all sections and subsections. (For example, this can be taken to be at least three documents against each of the CSO distributed across all sections such that in every section, there are at least three knowledge items available. This is only an indicative measure. The actual target will have to be set by manufacturing organization based on the priorities and significance of its corporate goals.) After 3 months of going live, the performance against each of the above KPIs has to be measured and refined. At this stage more specific targets based on the observations from the past months will have to set against each of the KPI. The KPIs have to be tracked regularly and reported every month. Further, the KPIs have to be refined every 6 months subsequently.

6.4.3 KM *Organization Structure Design*

KM acknowledges that the knowledge that exists within an organization's employees is its key to success; therefore, a shift from a technical/process focus to a more people-oriented focus is pertinent. The emphasis is not only on the processes that enable information to be provided and used effectively but also on the personal attributes necessary to take on the required facilitation and communication roles. For successful implementation and carry forward of the KM initiative, knowledge organization structure with well-laid out roles and responsibilities is absolutely essential for any organization. Taking into consideration the specific nature of operation of case study organization, the following knowledge organization structure is recommended for this case study.

The roles discussed in the proposed KM organization structure are:

1. Chief knowledge officer (CKO)
2. KM administrator
3. KM manager
4. KM analyst
5. KM engineer
6. KM expert
7. KM technical support

All the above roles are rotational roles having a minimum tenure of 18 months and a maximum tenure of 36 months. The KM organization structure is detailed in Fig. 6.11.

6.4.3.1 Roles and Responsibilities

The roles and responsibilities for the different designations as defined in the above KM organization chart are described below.

Roles and responsibilities of chief knowledge officer (CKO):

- Providing overall vision, broad policy direction, and guidance to the KM core team at case study organization.
- Overseeing the establishment, implementation, and evaluation of case study organization's KM program.
- Monitoring the KM activities of the offices, divisions, and units to ensure that the occupational priorities are addressed along with critical initiatives of knowledge throughout case study organization.
- Coordinating with case study organization KM manager to identify, select, assign, and evaluate the results of KM pilot activities.
- Forming and chairing the organization KM steering committee.
- Program managing all organization-wide KM activities.
- Implementing "rewards and incentives plan."
- Conducting an annual program review of results, lessons learned, and improvements in case study organization KM program activities.
- Measuring the impact of KM on the business.
- Defining case study organization KM policies, procedures, and guidelines.
- Budgeting and managing the funds allocated for organization-wide KM activities.
- CKO must think holistically and strategically and must be able to convincingly communicate the value of KM to skeptical audiences.

Roles and responsibilities of KM administrator:

- Supporting the conduct and evaluation of KM pilot activities.
- Developing, overseeing, and conducting training on all aspects of KM.
- Serving as the staff liaison between the case study organization KM community of practice and the KM steering committee.
- Monitoring "rewards and incentives plan."

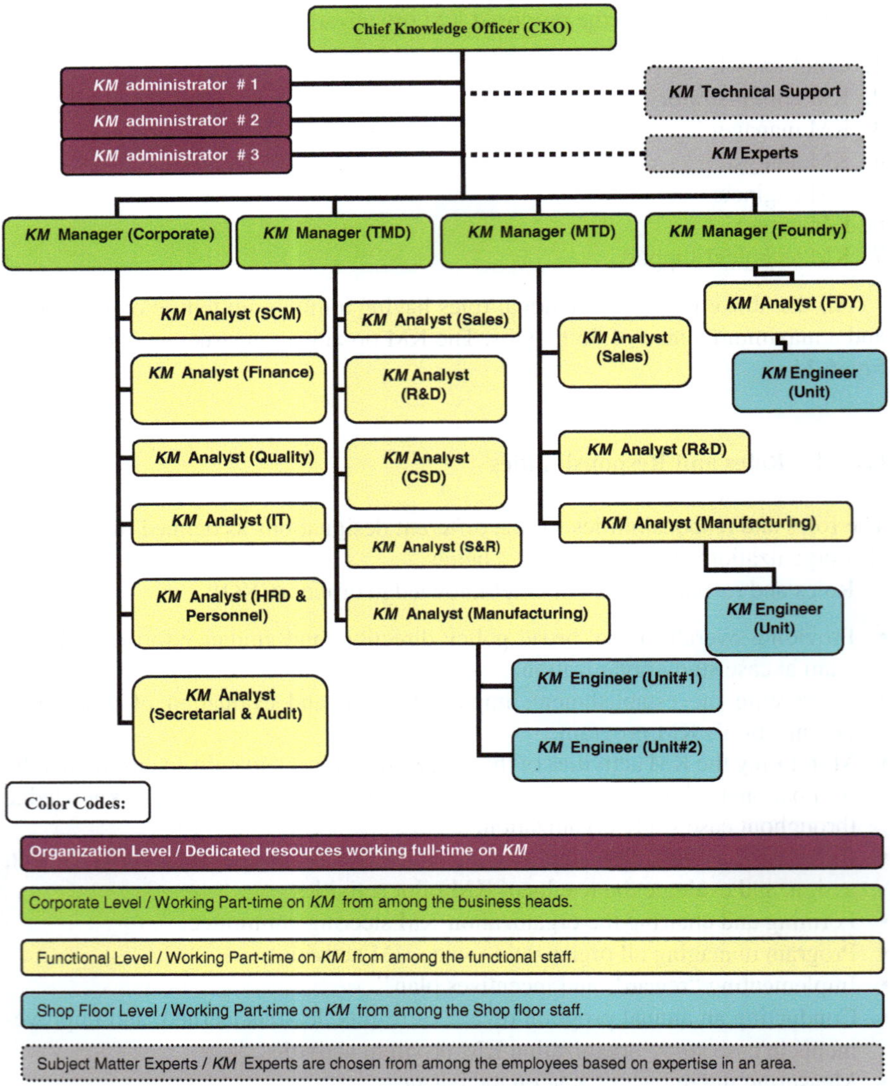

Fig. 6.11 KM organization structure

- Conducting focus groups and other appropriate activities to determine case study organization KM needs.
- Communication on KM pilot projects and development efforts to KM managers for incorporation in their respective areas.
- Maintaining and updating the case study organization KM portal. These include the following:

  - KM administrator would review the documents and upload it after assigning the required attributes, using the content management tools' content upload workflow.
  - Ensuring that the knowledge items are available as and when they are due:
    - Publish
    - Reject with reason
    - Change profile values
    - Change categories
  - Scheduling chat sessions with respect to requirements and invites users for the chat.
  - Creating discussion forums under particular domains and sub-domains. He can modify users list at any point of time. Create a new topic and also can approve/reject topic requests from users for discussion.
  - Maintaining list of domains as a master list.
  - In case there is a change in the taxonomy, KM administrator would need to move the contents accordingly:
    - Add new taxonomy items
    - Delete taxonomy items
    - Move contents from one head to another head
  - The KM administrator will have to continuously track the usage of the contents in KM and accordingly create or modify the domains/categories in the master list.
  - Housekeeping activities on the KM system such as delete old/irrelevant content.
- Facilitating the incentive plans for KM. KM administrator with the help of the administrator system should do the following tasks:
  - Generate report on the contributions and usage for the appropriate month/quarter and level of the organization and run the report.
  - Select the contributor/user of the month.
  - Plan and facilitate to provide the incentives for the respective outcomes.

Roles and responsibilities of KM manager:

- Ensuring the development and implementation of their divisional KM programs
- Leading development of KM strategies
- Selecting appropriate common techniques to implement those strategies
- Heightening awareness and interest in KM
- Communicating expectations for KM to the staff
- Clarifying communications to the staff and ensuring that the dissenting points are heard and understood
- Promoting inter-division and interunit knowledge sharing
- Sharing KM best practices and communicating them to their division
- Periodically providing HR with forecasts of KM gaps and proposed solutions

- Serving as members of the organization KM steering committee
- Linking KM to critical business processes and initiatives
- Reporting KM successes and risks to senior management
- Communicating lessons learned from KM activities in other divisions to those involved in KM development activities in their divisions
- Developing measures and metrics to monitor the effectiveness of divisional KM activities
- Providing feedback to the CKO regarding results and lessons learned in their KM initiatives

Roles and responsibilities of KM analyst:

- Participating as members of case study organization's KM community of practice
- Supporting the office or unit KM program and divisional KM champion
- Identifying budgeting needs and applying resources to develop and implement KM activities
- Conducting focus groups to obtain staff input on KM needs, capabilities, and techniques
- Promoting interunit knowledge sharing
- Sharing KM best practices and communicating them to their units
- Participating in and leading KM pilot activities to support identified priorities of their office or unit
- Providing feedback to the divisional KM champion regarding results and lessons learned in their KM initiatives
- Periodically meeting with employees to identify knowledge gaps (personnel and procedures) and preparation of action plans
- Special endeavor toward populating the KM portal by encouraging people to contribute and share
- Ensuring highest quality of document upload in terms of spelling, paragraph, font, title, relevant abstract, PDF conversion (wherever needed), etc.
- Keeping all the divisions/units updated on the contributions made by them, ensuring a healthy competition between the regions
- Deciding the contributor/user of the month
- Always keeping on the lookout for innovative ideas for knowledge creation, for example, through chats, discussion boards, personal interviews, or group discussions
- Designing the questionnaire for measurement survey of KM effectiveness
- Deciding the methodology for measurement whether it is manual or web based
- Helping units in collection of survey data
- Responsible for overall collection and monitoring of conducting the survey across the units
- Completing analysis and presentation of the findings and results

Roles and responsibilities of KM engineer:

- Ensuring that the staff/shop floor people understand and meet case study organization's expectations from KM

- Encouraging the shop floor people to share learning and knowledge
- Ensuring that the staff is trained on the use of common KM practices and techniques
- Ensuring that staffs receive appropriate rewards and recognition for knowledge sharing
- Ensuring that new and current employees are aware of the KM initiative and the importance of maintaining accurate and current information

Roles and responsibilities of KM expert:

- Providing domain/functional area focused inputs
- Participating as members of case study organization's KM community of practice
- Reviewing of various documents submitted in the KM portal
- Providing rating of the documents

Roles and responsibilities of KM user:

- Understanding and meeting case study organization's expectations from KM.
- Providing rating of the documents (user feedback).
- Communicating expectations from KM to the KM managers/KM analysts.
- Participating in chats, discussion forums, and online collaborations. For example, the user should request for a chat along with information on the topic and time for chat.
- Participating as members of case study organization's communities of practice.
- Participating in e-learning course modules.

Roles and responsibilities of KM contributor:

- Contributing knowledge items toward content creation for the KM portal
- Providing authentic and relevant information
- Complying with the metadata attribute requirements for any document
- Researching and preparing knowledge items useful toward meeting case study organization's learning, technological, and innovation needs
- Volunteering for mentoring programs

Roles and responsibilities of KM technical support:

- Providing KM technical support for the KM system
- Evaluating and improve the functionalities of the KM system over time
- Ensuring accessibility and availability of the KM system to the employees of the case study organization
- Resolution of technical issues related to the functioning of the KM system

6.4.4 KM *Reward Design with Incentive Plan and Matrices*

Very often the performance and motivation of employees is the most significant factor influencing his participation in any activity through which a company's revenue and profitability can be increased. This section describes the incentive plan and matrices for evaluating the participants in KM. The primary focus here is to highlight the prominent participants so as to encourage others to follow suit. Keeping in mind both the most basic aspects of KM, i.e., knowledge capture and knowledge use, the incentive plan has been described in twofold, the KM contributors and the KM users. In both the cases, the idea is to highlight the most prominent contributor and the most prominent user of KM.

The contributor of the month is the person who has contributed the maximum to the KM repository during the month. The contributor of the month is decided at three levels according to the rating considered for all the employees in the company, namely, the company level (whole of case study organization), the division level, and the unit level. The system should be able to capture the contributions of each individual and produce reports for KM administrator to facilitate their process of deciding the contributor of the month.

This is an activity designed to identify significant contributors to the KM repository every month. This is performed by the KM administrator. The report as described below is generated by the system, and KM administrator will analyze the report to decide the winners. The system will pick the concerned person image from a folder and will be displayed. The name of the image will be as "userid.jpg". The system should keep a record of the winners. The key elements of selection factors rating from expert, number of views or downloads by the users, and rating from the users. Each document/article will be measured on three parameters.

1. Coverage – Does the document cover the subject well?
2. Utility – Do you think this document is useful in the functional area?
3. Learning – Did this document teach you something new?

The relevancies are measured with ratings (1–3):

- 1 is satisfactory.
- 2 is good.
- 3 is excellent.

The users and expert will read the document and give a feedback on each document. The expert feedback will be captured while approving the documents. The user feedback will be captured through the following two ways:

1. On close of the document from the web browser, the system will generate an alert message saying that "Do you want to give a feedback." On selection of the "Yes" response, the system will open a feedback window for capturing the feedback; on selection of "NO," the system will cancel the operation.
2. There is another link called "Feedback" against each document; on click of this link, the system will open a feedback window to capture feedback on the corresponding document.

The latest rating given by user/expert would be considered for the contributor identification.

Rating calculation procedure: A contributor earns the average of the expert rating for the document (A). For example, if the expert gives the following ratings against each parameter for a document,

- Coverage – 3 (excellent)
- Utility – 1 (satisfactory)
- Learning – 2 (good)

then the average is = (3 + 1 + 2)/3 = 2.

A contributor earns points on the number of unique downloads or views on the document by a user for the first time (B). A contributor earns the average of user rating for the document (C). For example, if the two users give the following ratings against each parameter for a document,

User #1 rating:

- Coverage – 2 (good)
- Utility – 1 (satisfactory)
- Learning – 3 (excellent)

then the average for user #1 is = (2 + 1 + 3)/3 = 2.

User #2 rating:

- Coverage – 2 (good)
- Utility – 2 (good)
- Learning – 2 (good)

Then the average for user #2 is.......... (2 + 2 + 2)/3 = 2.

Then the consolidated user rating the contributor earns is

Average user rating = ((user #1 rating) + (user #2 rating))/No of users
= (2 + 2)/2 = 2

P = Max ({30 % of B} or {Number of documents rated/number of documents downloaded})

User rating (C) = (Average user rating) * (P)

Final rating for each document = A + B + C

Total rating for the Contributor = Final rating of Doc1 + Doc2 + Doc3.....etc. All inputs are consolidated from the users and experts and the system will generate a report as follows. The ratings shown on this table will be a simple average of the ratings obtained by the contributor for each of his contributions from the experts and the users. Sample table for the contributor is shown in Table 6.1.

Final rating column should be orderable ascending and descending. This report will be generated for the following options such as the company level (whole of case study organization), the division level, and the unit level. For each of the options, the report can be generated for the following time periods such as monthly, quarterly, and yearly. The contributor of the month must then be selected from the report and the selected person will be recorded in the system. The same contributor photo and

Table 6.1 Sample table for rating of the contributor

Contributor name	No. of contributions	Expert rating	User rating	No. of views/ downloads	Final rating	Select
Contributor #1						
Contributor #2						

Table 6.2 Sample table of the report format user contributions

S/no.	Title of document	Uploaded date	Expiry date	No. of views	No. of downloads	Document rating

month would be displayed in the home page. There will be a link "How does one become the Contributor of the Month" provided under the contributor of the month photo. On click of the link, the system will display a static page which contains Calculation procedure for Contributors in a separate window.

This is a link on the KM site that will be accessible to all the normal users. Under this link user can see his/her composite rating and leading contender for "Contributor of the Month" and the leading contender's rating.

The display message is as follows:

Your composite rating for contributor	[X]: 99.99
Your leading contender for contributor	[Y]: 99.99

"Under My Contributions," users can view his/her contributions and accessible status of his documents for the period of 3 months. The report format is shown in Table 6.2.

Each page where any article is opening should have a link to "Feedback." This should open up a form, which will capture the feedback for deciding the "Contributor of the Month." The feedback should be available in three forms:

- After opening a document for reading and closing it, the user will be prompted to enter a feedback; if he clicks YES, the feedback form opens up for capturing feedback.
- After every download, a mail should go to the user, requesting for feedback.
- Every document should have the feedback link; on click of the link, the system will open a feedback window for capturing the feedback.

The process environment design for the contributor of the month is detailed in Fig. 6.12.

The user of the month is the person who has made maximum use of the KM repository during the month. The user of the month could be decided at three levels, namely, the company level (whole of case study organization), the division

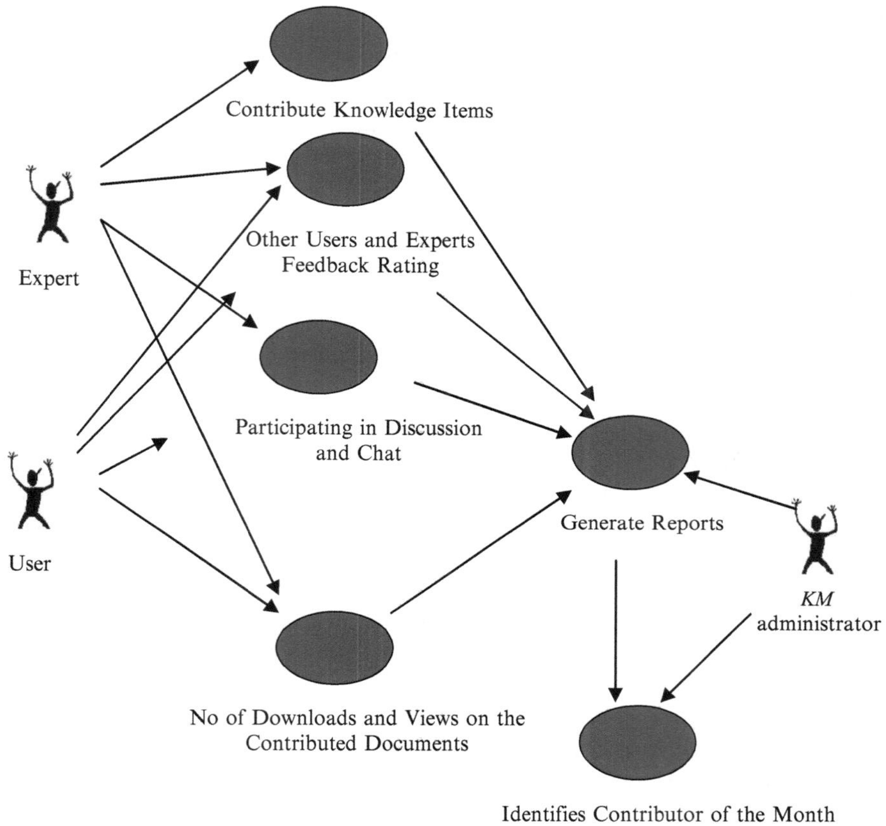

Fig. 6.12 Process environment design – contributor of the month

level, and the unit level according to the rating considered for all the employees in the company. The system should be able to capture the usage of each individual and produce reports for KM administrator to facilitate their process of deciding the user of the month. The following key elements will be used to decide the user of the month, namely, number of times the KM site was accessed, number of unique views or downloads on the documents, participation on chat, discussion with expert, and number of feedbacks given for the documents. The above key values are captured while the user is accessing the site. The above elements are measured as explained below:

- Access (A): Number of times the site was accessed by the user (maximum of one point per day for access irrespective of number of log-ins in a day into the system).
- Views/download (B): Number of document views/downloads from the site for the month (one point/view or download per document).

Table 6.3 Sample table of the rating report

User	Site access (A)	Views/ downloads (B)	Participation on chat (C)	Discussion with expert (D)	No. of feedbacks (E)	Rating (A+B+C+D+E)
User 1						
User 2						
User 3						

- Usage of chat (C): Each participation in the chat is to gain one point.
- Discussion forum (D): Each response for the month carries one point.
- Feedback (E): Number of feedbacks given by the user for the month.

Final rating for a user will be the sum of all the above for the use = (A + B + C + D + E). From the values for the above elements, the system will generate the report as follows in the Table 6.3.

Rating column should be orderable ascending and descending. This report will be generated for the following options such as the company level (whole of case study organization), the division level, and the unit level. For each of the options, the report can be generated for the following time periods, namely, monthly, quarterly, and yearly. The user of the month must then be selected from the report, and the selected person will be recorded in the system. The same user photo and month would be displayed in the home page. There will be a link "How does one become the User of the Month" provided under the User of the Month photo. On click of the link, the system will display a static page which contains calculation procedure for users in a separate window. The process environment design for user of the month is detailed in Fig. 6.13.

6.4.5 Communication Plan and Design

Communication about the KM initiative is one of the most important factors for the success of KM. Right communication would not only generate the much required awareness but also help in setting the expectation right with the employees. Repetitive and consistent communications via a broad set of channels is perceived as crucial for making the employees use and gain from the knowledge repositories. In this communication, theories about KM should be limited and focus should be on the success stories, anecdotes, and personal impressions of managers and employees. In the subsequent sections discuss the communication plan that can be used in manufacturing organization for its KM initiative. This plan has been discussed specific to the three phases of the KM solution rollout.

The three phases thus discussed are:

Phase 1 – Prelaunch
Phase 2 – Launch
Phase 3 – Postlaunch

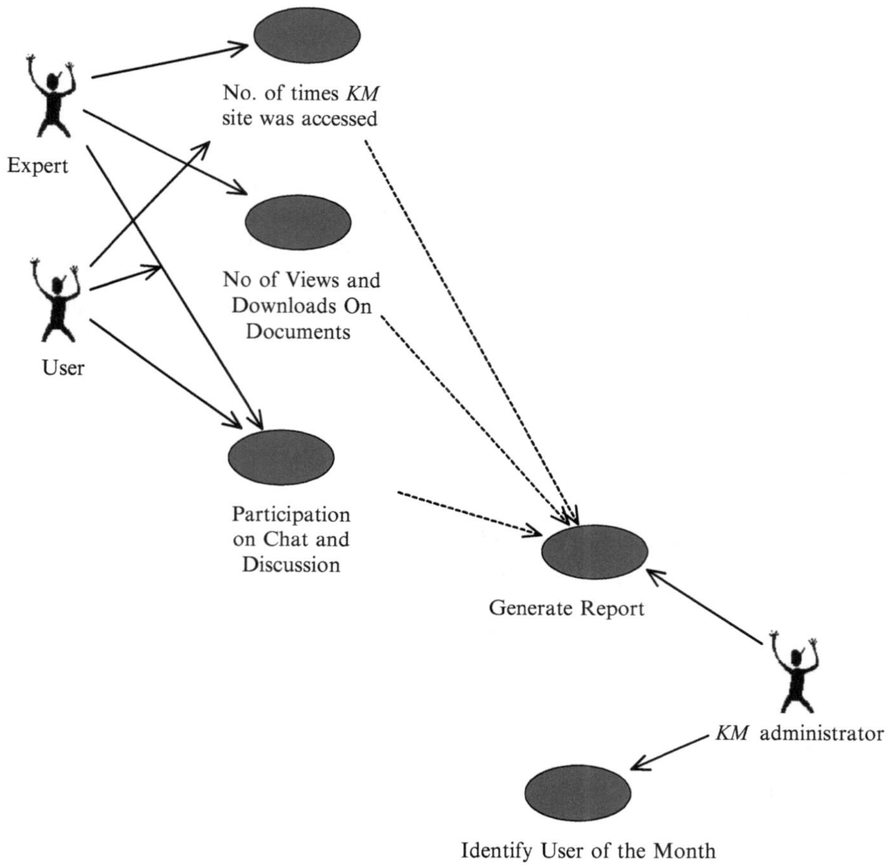

Fig. 6.13 Process environment design – user of the month

Phase 1 – Prelaunch

The communication objectives for the prelaunch phase are to generate anticipation and build momentum for the launch of the KM and to project the usefulness and benefits of KM.

Channels and activities: The various channels that can be used for promotion and awareness generation can be:

- E-mails
  - (a) Teaser mails: An advance or teaser mail to intimate the employees of the impending arrival of KM
  - (b) Info mail: An info mail with all basic details of KM (concept, framework, possible benefits, etc.)

Promotional posters/banners/brochures

- Creation of groups (The KM team has to be instrumental in spreading the news/word of mouth about KM and generate as much momentum as possible.)
- The employees are to be provided with an opportunity to come forward and ask questions about KM. A body in the form of "just ask us" can be formed, which would be responsible for answering all questions. Questions (along with answers) which are genuine and can be helpful in bringing clarity should be appreciated and acknowledged by publishing in the internal magazine of the organization.

Phase 2 – Launch

The communication objectives for the launch phase are to develop urge of the employees to visit the KM site and to provide education on KM site and its features. The basic objective should be to entice the users to make a visit to the KM portal. It is very important to launch with entries already written and published. That way, you allow users to come and explore the site which helps them decide right then if they wish to make a return visit.

Channels and activities: The various channels that can be used in this phase are:

- E-mail
    (a) Everyday for 7 days prior to launch
    (b) Education mails with proper information on the manufacturing organization's KM initiative, expectation from employees, roles and responsibilities, etc.
- Education campaigns
    (a) Navigation aspects of KM, etc.
    (b) Instruction/procedure for chat/discussion forums
    (c) Instruction/procedure for document upload/download
- User group presentations
    (a) Showcase successful KM initiatives and benefits derived out of it

Phase 3 – Postlaunch
The communication objectives for the postlaunch phase are to sustain the enthusiasm, to entice the users to visit and revisit the KM site, and to introduce new features in a periodical manner and keep the momentum up and going.

Channels and activities: The various channels that can be used in this phase are:

- Contests
  - (a) Quiz (on KM portal)
  - (b) Treasure hunt (locate a document from KM portal based on hints given at various places within the portal)
- KM web page
  - (a) New feature addition in the "What's New" section
- Status reports
  - (a) Weekly usage report (document download status for each division)
  - (b) Weekly contribution report (document upload status for each division)
  - (c) Weekly highlights (number of hits from each division, unique visitors, value packs released, etc.)
  - (d) Graphical representation in a way which can be easily understood by all
- Rewards and recognition
  - (a) Star user of the month
  - (b) Star contributor of the month

6.5 Summary

Implementation of KM involves the identification and analysis of available and required knowledge assets and knowledge asset-related processes and the subsequent planning and control of actions to develop both the assets and the processes so as to fulfill organizational objectives. At the strategic level the organization needs to be able to analyze and plan its business in terms of the knowledge it currently has and the knowledge it needs for future business processes. At the operational level, the process design is very critical and it forms the backbone for the implementation of KM solution. Devised generic process design framework can be used as a basis for any manufacturing organization, and from there the process design can be improved further according to the needs and objectives of an individual organization. KM is a continuous process and not a one-time activity. It is thus very essential that organization follows a framework of activities as part of its KM process. Also knowledge sharing is one of the most critical steps in KM activities. To achieve effective knowledge sharing, it is important to encourage workers to share their knowledge for the best interests of the firm. The generic process design, organization structure, KPI design, reward design, and communication plan design developed in this research can be directly taken as base for any manufacturing industry in building KM solution.

Chapter 7
Learning Design for Knowledge Management Solution Implementation

7.1 Introduction

The *KM*, organizational learning (OL), and learning organization (LO) have existed over the years as largely separate fields. There is a link between OL, LO, and *KM*, even though OL and LO primarily deal with learning (at the individual, group, and organizational levels), and *KM* deals with the management of existing and newly created knowledge within the organization. New knowledge is created when learning occurs at the individual level, and this learning must then be translated to the group and organizational levels, resulting in a new justified belief system for the organization.

7.1.1 *Organizational Learning*

Organizational learning (OL) can be described as the study of learning processes within the organization. OL is a process based on individual learning wherein organizations engaged in creating and obtaining knowledge can adapt to the changing conditions of the environment or to change the environment proactively. OL deals with the sociopsychological process of learning, and this field is largely descriptive in nature (Friedman et al. 2005; Örtenblad 2001; Sun and Scott 2003; Tsang 1997). This is captured in the definition offered by Sun and Scott (2003) wherein OL is described as "the learning process used in the organization. It deals with the question of how individuals in the organization learn." The literature discusses about OL as an important perspective in an organization for achieving successful management in the new economy (Levitt and March 1988; Prietula and Simon 1989; Behling and Eckel 1991; Isaacs 1993; Nonaka 1994; Crossan et al. 1999; Vera and Crossan 2004; Argyris 2004).

K. Ganesh et al., *Design and Development of Knowledge Management for Manufacturing: Framework, Solution and Strategy*, DOI 10.1007/978-3-319-02892-7_7,

7.1.2 Learning Organization

LO is primarily the domain of practitioners whose perspective centers on the characteristics of an organization that promotes learning and facilitates the creation of a certain type of organization (Tsang 1997) and deals with prescription necessary to create a form of an organization that is continuously capable of recreating itself (Sun and Scott 2003; Senge 1990). LO is thus considered more normative in nature. Although some researchers still use the terms OL and LO interchangeably (Crossan et al. 1999; Rahim 2002), the normative nature of the LO discipline is well established and is distinguished from the descriptive nature of OL (Robinson 2001). Therefore, given that an LO must be capable of continuously recreating itself, the type of learning capability is considered more double loop. In this research, for the purpose of discussing the linkage between OL, LO, and *KM*, the work of Senge (1990), which formulates five disciplines of the LO, will be considered. The five disciplines are team learning, personal mastery, shared vision, systems thinking, and mental models. These five disciplines inherently denote the levels of learning: personal mastery and mental models at the individual level, team learning and shared vision at the group level, while systems' thinking at the organizational level (Isaacs 1993; Argyris 2004).

7.1.3 Knowledge Management

Knowledge is a relationship between the knower and the known and only learners create knowledge. Knowledge is said to be the justified true belief of individuals (Nonaka 1994). More specifically, practices that arise out of the belief system must not create mismatches between actual and expected outcomes and only then true knowledge is transferred. The immediate purpose of *KM* is not to improve either worker effectiveness or an organization's bottom line. Its purpose is to enhance knowledge processing held at different levels (i.e., individual, group, and organizational levels). It is also imperative to note that *KM* does not directly manage, create, or integrate most knowledge outcomes in organizations but only impacts knowledge processes (performed by operational process agents), which, in turn, impact knowledge outcomes. Researchers have shown that lack of proper *KM* in an organization can reveal negative indicators of organizational performance (Bontis et al. 2002). Researchers have studied a three-tier nature of *KM* in the organization (Argote et al. 2003): creating or developing new knowledge, retaining the knowledge, and transferring knowledge. Stemming from this three tiers, computational view and organic view (Hazlett et al. 2005), or 1st-, 2nd-, and 3rd-generation *KM* (Cavaleri 2004), have evolved. Seely-Brown and Duguid (1991), Hazlett et al. (2005), Argote (2005), and Dyck et al. (2005) have researched and found that *KM* is inevitable for organizational management and organizational performance once knowledge transformation happens. This goes hand in hand with the "theory in action" theme of LO, where all knowledge has to be generalized and crafted in ways in which employees can use it to make the knowledge actionable.

7.2 Learning Culture

It is only recently that researchers have begun to make genuinely systematic connections between learning, knowledge sharing, and culture within organizations. A learning culture is a nonnegotiable criterion for the success of *KM* in any organization. The learning culture in an organization makes sure that there is a collaborative environment of knowledge transfer in the organization and that knowledge is extracted from distinct silos and used as the common knowledge of organization. Secondly, an organization should foster expansive learning environments by realigning its learning goals toward absorbing external knowledge and integrating it with internal knowledge to develop new insights. This concept of learning goal of the organization is found effective for inducting new roles and responsibilities, succession planning, knowledge transfer, and new initiatives for business/operational improvements (innovation) and to meet employee aspirations. As an initial step toward attaining the learning goal of any organization, it is important to mentor the employees to better their roles, responsibilities, and competencies; provide seminars, discussions, and chats by senior/experienced employees on their core competency areas; facilitate problems and solution sharing between employees; increase regular training on functional areas to upgrade employee skills; and facilitate and improve the accessibility to e-libraries, research databases, industry best practices, etc. These practices of an organization would improve the knowledge transfer and foster an environment conducive for *KM* initiatives through OL and LO.

7.3 Research Gap Based on Literature

Cumulative evidence from past research in *KM* suggests that effective implementation of *KM* solution in any organization requires robust learning designs and models for various critical elements of process, people, and technology. The primary intention of learning design is to devise and design the learning factors for the implementation of *KM* solution. Research and studies have also highlighted the need for organizational learning and learning organizations in a conducive learning organizational culture as the backdrop of effective *KM* implementing strategies. There are very few research efforts to map out these dimensions and its influence on *KM* initiation and success. This research therefore attempts to unfold the effects of improvement in the performance indicators of the organization from the implementation and prognosis of learning factors and *KM* initiatives.

A learning organization is a collective undertaking wherein organization, people, knowledge, and technology drive learning. These subsystems must work in harmony to achieve better and faster learning and increase an organization's relevance. Knowledge capturing and storing is easy when there is an efficient learning design and system, continually building and improving organizational practice. The learning design methodology influences the way people gather and interpret knowledge

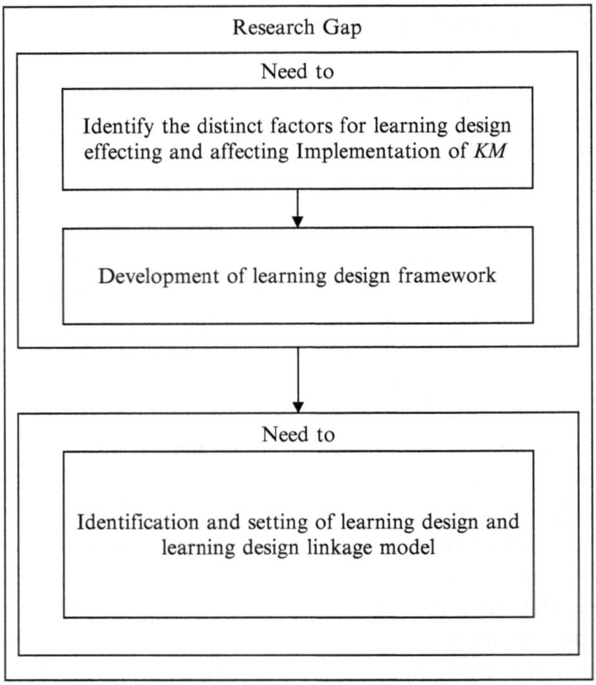

Fig. 7.1 Research gap

to make design decisions. There are several methods to capture learning design knowledge, one of which is the instructional design approach. The first step is to create a learning environment for the individual to acquire new knowledge and elements and thereby to generate processes of analysis. The next step is to bring learning activities which are designed to make employees achieve a given set of actions in order to help them internalize knowledge. This learning design falls in the category of "learning by doing" rather than just following the rules and doing. The development of learning methods then could focus on constructing e-learning activities from *KM* systems which is absolutely necessary to manage the complexity of e-learning issues in companies. Learning design can be reused in e-learning, and this has emerged as one of the most significant recent developments in *KM* implementations. Based on the detailed literature study, the research gap in the area of OL, LO, and *KM* is identified, and it is detailed in Fig. 7.1.

7.4 Research Process and Methodology

The development of learning design and identification of impact of learning factors for implementation of *KM* solution are detailed through this study. Research design is detailed in Fig. 7.2.

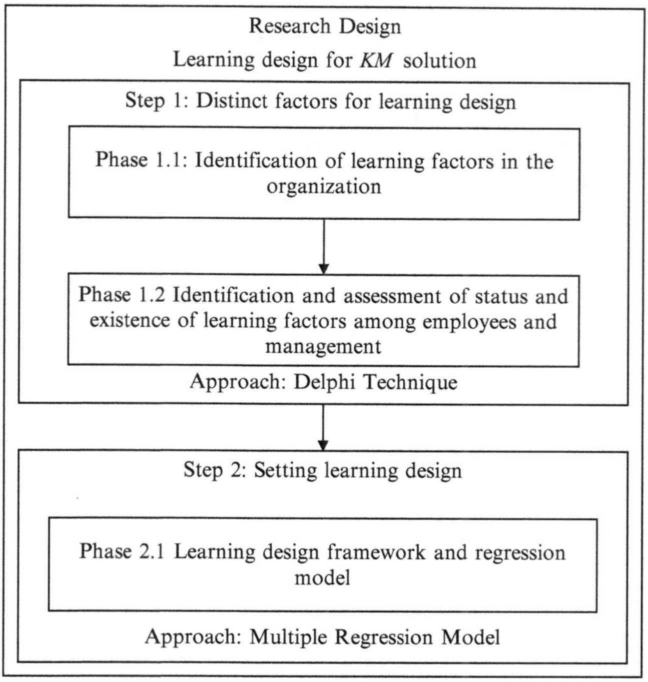

Fig. 7.2 Research design

7.4.1 Delphi Analysis

Delphi analysis was used to bring out the learning factors from the research and business literature. The learning designs with respect to the above goals and objectives are explained in detail in the research based on a Delphi analysis. The Delphi technique was also used to arrive at the conceptual framework of this research.

The learning designs based on the outcome of Delphi analysis are:

- Organizational learning
- Expert management system
- Communities of practice
- Mentoring
- E-learning

7.4.2 The Conceptual Model

From the Delphi technique it could be understood that *KM* is easy only with an efficient system to capture and store the knowledge, which forms the basis of

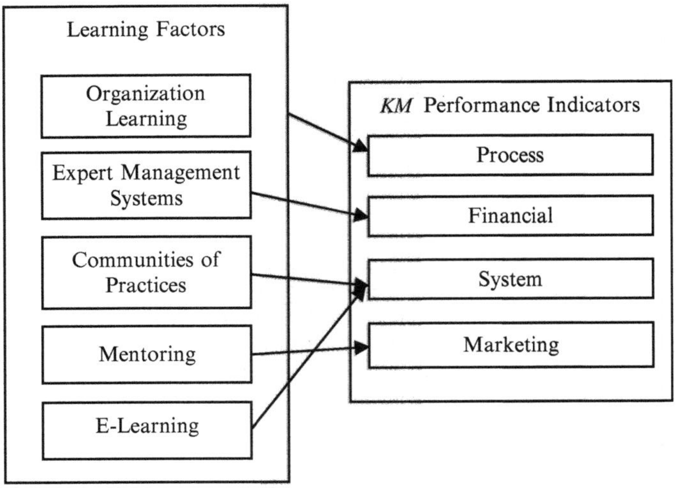

Fig. 7.3 Conceptual framework

learning design. There are several methods to capture learning design knowledge, and an important one is the instructional design approach. A learning environment with the basis of a constructivist model of learning theory, where learning is regarded as an active process in which meaning is developed on the basis of experience, should be created. Learning occurs only when it is situated in realistic settings where experience should be integrated with the task and not as a separate activity, which brings out the "learning by doing" paradigm. As mentioned earlier, e-learning technology is one that compliments the successful initiation and management of *KM* initiatives. The effects of learning factors may influence the improvement of performance indicators of *KM* directly or indirectly. The improvement in performance indicators of *KM* will have a positive impact on the achievement of strategic vision of the organization. This underpins the conceptual framework for this study that is depicted in Fig. 7.3.

7.4.3 Construct Definitions

The definitions of constructs are provided in detail.

7.4.3.1 Organizational Learning

This is a process to share knowledge at the organization level. The knowledge that is shared could be about news initiatives, individual's views and opinions on a subject, and a common forum for globally addressing a problem or for collection feedback on success and failures within the organization or external environment.

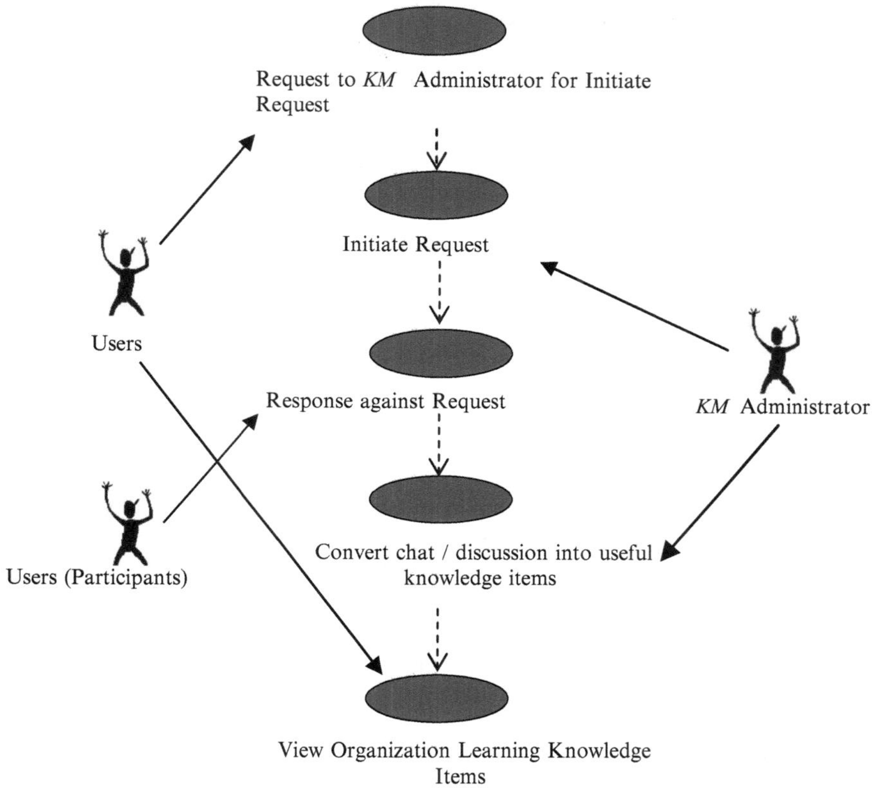

Fig. 7.4 Process environment

The *KM* administrator would post a knowledge-sharing request (directly or on request of a user) on the site under the section called "Organization Learning." The system could capture the following information (Fig. 7.4):

- Objective of the request
- Community (user/user group to participate)
- Mode of communication (chat/discussion forum)
- Date and time
- Expected learning

Communities (groups of people who share similar goals and interests) are created based on the learning sought. These could be temporary (created only for a certain event) or relatively permanent. The *KM* administrator would assign groups of people going to participate in the discussion. Those selected people will have rights to contribute to the organization learning contributory session. The same group of people only gets mail or alert from the *KM* administrator.

The mode of communication could be either chat/discussion forum or both. The thread would be stored by the system and would need to be converted into useful knowledge items by *KM* administrator (externally and manually). The entire transcript would be available under archives and catalogued for easy retrieval. The conversion of the chat transcript/threaded discussion into a knowledge document will be done using the functionalities already available in those respective sections.

Key roles:

- *KM* administrator
- Participant
- User

7.4.3.2 Expert Management System

This section is a forum for users within manufacturing organization to locate experts within the organization and get inputs from them.

A user can reach the expert's page through a link on *KM* site, which will list down all the experts and their areas of expertise. Alternatively, a simple/advanced search on a specific topic can lead to the expert's page in the specified areas. The expert's page would display all relevant information about the expert like name, division he belongs to, contact, and other relevant links like expert opinion, FAQs, tips, and suggestions.

7.4.3.3 Communities of Practice

A community of practice is a group of peers with a common sense of purpose who agree to work together to share information, build knowledge, develop expertise, and solve problems. Communities of practice are characterized by the willing participation of members and their ongoing interaction in developing a chosen area of practice.

In other words, communities of practice are learning forums where members teach and learn from each other and use each other as a sounding board. Communities of practice may focus on problem solving, knowledge sharing, and innovation. Each knowledge community can further consist of certain subcommunities which attempt to focus on smaller functional areas.

Guideline to establish and facilitate CoP at manufacturing organization:

These guidelines provide tips on establishing and facilitating communities of practice; they are not intended to be prescriptive, as communities of practice may vary according to their particular purpose and membership. The guidelines aim to support the facilitation and establishment of communities of practice, provide a consistent high-quality experience for participants, and thereby contribute to optimizing the outcomes for community of practice members.

Step 1: Identification of member needs – An e-mail survey may be a useful tool to get detail about the needs of potential members and scope the parameters of the community of practice. A suggestive questionnaire can be in line of the following:

- What are your three key information and learning needs?
- What activities is your area currently undertaking in relation to the above areas?
- What do you expect to gain from participating in the community of practice?
- Are you interested in being part of a small informal steering group that would assist with devising the forward program of meetings and activities?
- We anticipate holding presentations at some meetings – who do you suggest could be suitable presenters and on which topics? (Note: this could include you.)
- What are your preferences with regard to meeting times, intervals, etc.?
- Based on your experience, what are the key success factors for this community of practice?

The survey results are to be analyzed with emphasis on the following:

- The level of demand and need for knowledge and capacity building in the particular topic area
- The willingness and capacity of potential members to contribute to supporting the group
- The level of knowledge and expertise of members in relation to the topic and their willingness to share this with each other

Step 2: Business case – A robust business value has to be ensured in terms of:

- Degree of alignment of the topic with manufacturing organization's business objectives
- Level of priority of the topic among potential CoP members
- Whether there is a preexisting community of practice or forum in the topic
- The available capacity within manufacturing organization to provide facilitation support, especially in the initial phase of establishment
- Indicative value contribution

Step 3: CoP team structure and operating principle – The CoP team should be given the form of an official entity to track the workings and outputs. The structure should ideally be in terms of:

- Sponsorship – high-level advocate and stakeholder
- Mission and objectives – the who, what, and why
- Roles – leader, SME, knowledge expert, core team, etc.

So as to ensure that the process of identifying goals, terms of reference, and operating principles encourage members to own the community of practice, the following points may be useful to take into account:

- If a survey is carried out of potential members' goals and expectations, the results can be presented at the first meeting for discussion and agreement.
- Once agreed, goals and terms of reference can be published on a web page and be periodically reviewed in consultation with members.

- Operating principles can also help members clarify their expectations of each other.

The rules set for CoP operation should be practical and inspirational, for example:

- At least one new learning from each meeting.
- Privacy and confidentiality is maintained within the community.
- Views expressed are those of individual practitioner members.
- After each meeting a summary of the discussion is circulated to members and uploaded to the *KM* portal.

Step 4: Identification of scope of *KM* team and online support – The *KM* team has to act as a facilitating group and encourage cross-division ownership.

In addition the *KM* team would be entrusted with the following:

- Regular meetings to plan the forward program and identify how facilitation can be shared.
- The team should help identify expertise, resources and references, presenters, site visits, venues, and topics for the broader community, as well as facilitate meetings.
- Consider establishing an e-mail discussion list to help communication flow and facilitate relationships across the team and participants.
- Ensure support of the CoPs from relevant units/divisions.
- Encourage and facilitate the CoPs to become self-supporting.
- Ensure that information sharing can happen from wherever the expertise lies, including within the group, from non-members and/or other units/divisions.
- An e-mail discussion group may be set up to encourage member's engagement and share expertise and information.
- Links to information about communities of practice can be made from the communities of practice web page.
- Technical support and facilitation for setting up of discussion forums for the CoPs.

Step 5: Evaluation and measurement – Seeking regular feedback from members and periodically evaluating outcomes can be a useful means of measuring the "health" and relevance of the community of practice. Emerging issues can also be identified through these processes. Timing evaluation to feed into the planning cycle can assist with identifying its future. The following points could be included in the evaluation:

- Level of participation in e-mail discussion, presentations, and meetings
- Range of agencies involved
- Attendance at meetings
- Outputs achieved, such as better practice checklists and toolkits
- Evaluation of the uptake and usage of these checklists and toolkits
- Member satisfaction

Step 6: Closing a CoP – A community of practice may be closed in any of the following circumstances:

- The group is no longer active.
- It has achieved its principle purpose.
- It has been assessed by the lead agency as no longer serving its original purpose, is no longer considered to be an organizational priority, or has drifted from its agreed mission.
- It has failed to become self-supporting.

Suitable processes for closing a community of practice may need to be taken into account. Issues to consider include:

- Consulting members regarding closing the community of practice
- Recognizing the group's achievements
- Acknowledging member's contributions
- Notifying members that the community of practice is closed, via e-mail and at meetings as appropriate

A joint effort by the *KM* team and the CoP members to summarize the knowledge gained, practical tips, and lessons learned for future projects in the form of K-products has to take place at the closing of all CoPs.

CoPs relevant to manufacturing organization

Some of the CoPs which can be set up for different areas during the initial phase are:

- CoP-new product development: This community of practice would provide a mechanism for sharing information on various aspects of new product development like product engineering, market analysis, product lifecycle management, and time to market.
- CoP-emerging technologies: This community of practice would provide a mechanism for sharing information on emerging technologies in the textile industry and the feasibility of adopting these technologies at manufacturing organization.
- CoP-heat treatment: This community of practice would provide a mechanism for sharing information on various manufacturing techniques used to alter the hardness and toughness of a material, mostly metallurgical. The techniques can include annealing, case hardening, induction hardening, precipitation strengthening, tempering, quenching, and the probable effects on the materials used in manufacturing organization.
- CoP-lean manufacturing: A forum where individuals can share variety of perspectives on potential areas of waste elimination and how to reduce the seven wastes and its effect on production time, delivery, quality, and cost.
- CoP-innovation: CoPs can act as breeding grounds or innovation. Some communities of practice are based around technical or professional peer groups, and the focus is on new and emerging areas of knowledge where there may be the opportunity to gain an edge. These groups can provide the basis for rapid dissemination of new ideas or products.

Table 7.1 Suggested e-learning courses

Target audience	E-learning modules
Top management	Negotiation skills, business etiquette, managing change, BSC, interpersonal and leadership skills, lean manufacturing, motivational techniques
Middle management	Presentation skills, performance management, coaching skills, mentoring skills, six sigma, BSC, interpersonal and leadership skills, people management, problem-solving techniques, waste elimination, inventory management
Lower management	BSC, sales tips, basics of IT programming, etc.
General	Managing time at work, wealth management, health tips, investment tips, retirement planning, etc.

7.4.3.4 Mentoring

Mentoring is a one-to-one caring, supportive relationship between a mentor and a mentee that is based on trust. The mentor is simply a wise and trusted friend with a commitment to provide guidance and support for the mentee to develop their fullest potential based on their vision for the future.

Mentoring and coaching have become very popular methods of training and knowledge transfer in recent years. By matching new or inexperienced employees with more experienced senior personnel, the intangible, tacit knowledge of manufacturing organization can be passed on effectively. It allows the new employees to grow without learning the hard way and creates a bond between mentor/coach and mentee. Mentoring and coaching also allow the more experienced personnel to "give back" to the organization.

7.4.3.5 E-Learning

E-learning is the delivery of a learning, training, or education program by electronic means. E-learning involves the use of a computer or an electronic device (e.g., a mobile phone) in some way to provide training, educational, or learning material. A shift to e-learning is a shift toward a new learning culture. An effective e-learning strategy will take the long view and be built on a strong business case. It will range from technology issues to environmental factors and will consider the transition from a change management perspective. One has to have a solid understanding of the needs and the plethora of technology-enabled learning options to lay the foundation for both short-term and long-term success in changing the learning culture of manufacturing organization. Some of the suggested e-learning courses that manufacturing organization needs are indicated in Table 7.1

7.4.3.6 Process Performance Indicators

The performance improvements in relational exchanges tend to be measured from a user perspective. As such, the present study will adopt a user perspective with respect to performance improvements. Examples of performance improvements among users can include, but are not limited to, lower searching costs and average cycle (Lieb and Bentz 2005).

7.4.3.7 Financial Performance Indicators

Financial performance is defined by the financial and market measures to evaluate the companies' efficiency and effectiveness, such as improvement in search speed and cost reduction (Zhao et al. 2006).

7.4.3.8 Strategic Performance Indicators

Brown et al. (2007) found that manufacturing managers were more involved in the strategic planning process in world-class than non-world-class plants, reflecting the importance of functional involvement in improving performance. Strategic performance relates to a company's performance in serving customers in terms of cost competitiveness (Hayes and Wheelwright 1984; Roth and Miller 1990) and service and quality excellence.

7.4.3.9 Marketing Performance Indicators

It is important to measure marketing performance to reflect the changing environments and strategies. Many researchers have measured marketing performance in terms of different operational or financial measures, such as profit margins (Bititci 2000).

7.4.4 Hypotheses

In the context of *KM*, a critical requirement is that the learning factors develop relationships in a coordinated and collaborative way in order to improve performance indicators. This study focuses on the following four research hypotheses:

H1. There is a significant impact of organizational learning on process performance indicator.

H2. There is a significant impact of expert management system on financial performance indicators.

H3. There is a significant impact of communities of practices and e-learning on the system performance indicators.
H4. There is a significant impact of mentoring on the marketing performance indicators.

7.4.5 Research Methodology

The survey constructs used for this study were derived and developed based on the detailed literature review and from semistructured interviews with *KM* managers, practitioners, and executives of leading manufacturing firms in India. All the questions in the survey are collected on a five-point Likert scale. The reason for choosing a five-point Likert scale is to maintain the reliability of respondents' perceptions (Lissitz and Green 1975).

7.4.5.1 Sampling and Data Collection

To test the hypotheses the relevant data were collected from manufacturing firms in India. A questionnaire survey methodology is chosen for the empirical survey, since it is believed to be the effective way of attaining a large number of manufacturing firms. The questionnaire is designed as close-ended questions so that executives of companies can answer via e-mail. Sample manufacturing firms were identified from the business directory of the India Trade and Development which was the resource center for all business databases and information of industrial companies. A hard copy survey kit including questionnaire with a self-addressed envelope, a statement about the purpose of research, confidentiality agreement, and statement and anonymity agreement was given to the respondents. The survey kit also includes an assured statement for sharing of results and implications after the completion of empirical study. All of the above were included in the soft copy survey kit also. Three follow-up reminders were provided for both hard and soft copy survey.

A total of 150 questionnaires were sent to manufacturing firms. After three follow-ups, 85 useable responses were received, representing a response rate of 57 %, which compares favorably well to other such empirical studies. The manufacturing firms involved in this survey is practicing and following *KM* from minimum of 1 to maximum of 2 years. The proportion of manufacturing firms with number of years is indicated in Fig. 7.5.

7.4.5.2 Construct Testing

During a pretest, *KM* executives were asked to point out the degree to which certain measures tap the construct domain and to recommend modifications.

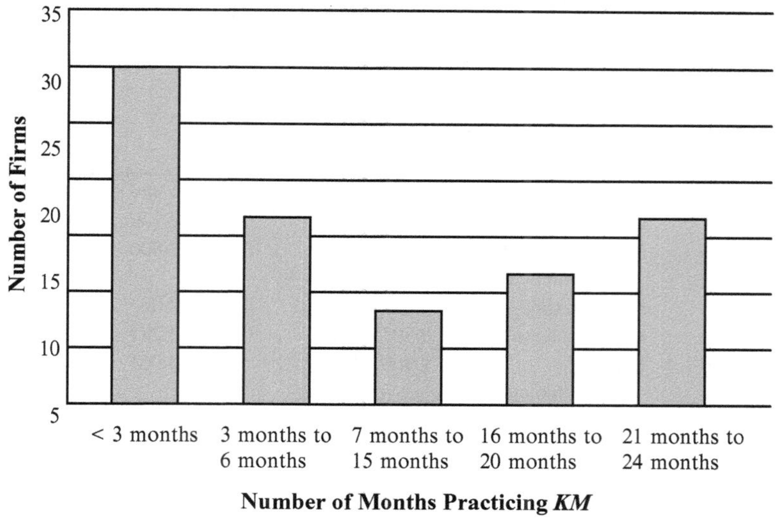

Fig. 7.5 Proportion of firms with number of months practicing *KM*

7.4.5.3 Reliability Testing

Reliability (internal consistency) of the items comprising each dimension was examined using Cronbach's alpha (Cronbach 1951). Following the guideline established by Nunnally (1978), an alpha score of higher than 0.60 is generally considered to be acceptable, whereas an alpha score of higher than 0.80 is considered a good measure of reliability. Cronbach's alpha was used to measure the reliability of the hypothesized individual constructs. A commonly used value for acceptable reliability is 0.70 (Hair et al. 1998). More reliable measures give greater confidence that the individual indicators are all consistent in their measurements, and therefore, the model is repeatable. The Cronbach's alpha scores for the constructs of this study like organization learning, process performance indicators, expert management system, financial performance indicators, communities of practices, e-learning, system performance indicators, mentoring, and marketing performance indicators are 0.878, 0.825, 0.821, 0.921, 0.880, 0.834, 0.856, and 0.818, respectively. All alpha scores are higher than 0.80 and thus depict a good measure of reliability.

7.5 Empirical Analysis and Results

The results were then analyzed using the statistical software analysis package Statistical Package for Social Sciences (SPSS). A multiple regression analysis is a statistical technique that allows us to predict a respondent's score on one variable on the basis of their scores on several other variables. A multiple linear regression model is an extension of a simple linear regression model to incorporate two or

Table 7.2 Contribution of organization learning to improvements in process performance indicators

Dependent outcome	Independent measure that has significant partial regression coefficients	B	Std. error	Beta	*t*	*p*	*Adj* R^2; *P*
Lower searching costs	Objective of request	1.857	0.310		5.984	0.000	0.509
	Community	0.380	0.178	0.393	2.139	0.034	0.000
	Mode of communication	0.223	0.175	0.226	2.800	0.006	
	Date and time	0.335	0.161	0.345	2.075	0.039	
	Expected learning	0.303	0.132	0.276	2.287	0.023	
Average cycle length	Mode of communication	1.960	0.640		3.061	0.002	0.236
	Date and time	0.303	0.119	0.261	2.540	0.012	0.000
Service level improvements	Objective of request	2.063	0.628		0.329	0.001	0.220
	Community	0.276	0.117	0.241	2.362	0.019	0.001
	Mode of communication	0.249	0.126	0.263	1.970	0.050	
	Expected learning	0.303	0.119	0.261	2.443	0.016	

more explanatory variable in a prediction equation and to identify its effect on the response variable. Multiple regression modeling is now a mainstay of statistical analysis in most fields because of its power and flexibility. Multiple linear regression was used to test the proportion of variance in the chosen outcomes explained by the following groups of independent variables or measures: organizational learning (H1), expert management system (H2), communities of practices and e-learning (H3), and mentoring (H4). Executives were asked to rate their responses to these variables on a 1 (least important) to 5 (most important) Likert scale for those items in order to test these hypotheses.

7.5.1 Survey Findings

H1. There is significant impact of organizational learning and process performance indicators.

In order to examine the relationship and extent of effect between the independent variables and dependent variables, the multiple linear regression analysis has been carried out for the data collected. The independent variables/measures that represent the organization learning are objective of request, community, mode of communication, date and time, and expected learning (X_1,X_2.....X_5). The dependent variables/outcomes representing the process performance indicators are lower searching costs and average cycle length and service level improvements (Y function). The linear regression model is given by $Y = a_0 + a_1X_1 + a_2X_2 + \ldots\ldots + a_5X_5$.

Table 7.2 shows results of multiple linear regression analysis including the level of significance for the relationship between the dependent outcomes and

Table 7.3 Contribution of expert management system to improvements in financial performance indicators

Dependent outcome	Independent measure that has significant partial regression coefficients	B	Std. error	Beta	*t*	*p*	*Adj* R^2; *P*
Improvement in search speed	To locate experts and get inputs	4.432	0.470		9.440	0.000	0.474
	User can reach the experts page	0.301	0.113	0.266	2.585	0.011	0.000
	Experts page display all relevant information of experts	0.301	0.130	0.271	2.116	0.029	
Cost reduction	To locate experts and get inputs	4.307	0.495	8.709	3.283	0.000	0.173
	User can reach the experts page	0.173	0.087	0.191	1.987	0.049	0.007
	Experts page display all relevant information of experts	0.308	0.078	0.327	3.952	0.000	

independent variables/measures and also the *t* statistic values of those independent variables which have significant partial regression coefficients. Table 7.2 reveals that the tested regression models are meaningful ($p<0.05$), leading to the conclusion that the organizational learning can contribute significantly to the improvement of the process performance indicators.

The results presented in Table 7.2 indicate the support for H1. The contribution of organization learning was significant for dependent variables/outcomes as revealed by the values of $p<0.05$.

H2. There is significant impact of expert management system on financial performance indicators.

The expert management system was identified, and these are the independent variables/measures of H2. The items considered for H2 are as follows: to locate experts and get inputs, user can reach the experts page, experts page display all relevant information of experts. The dependent variables/outcomes are improvement in search speed and cost reduction. Table 7.3 shows the results of the analysis that supports the hypothesis that the expert management system contributes to the improvement of the financial performance indicators, as the tested regression models are observed to be meaningful ($p<0.05$).

The results presented in Table 7.3 indicate the support for H2. The contribution of the independent measures was observed to be significant for five dependent outcomes with a level of significance, $p<0.05$.

H3. There is significant impact of communities of practices and e-learning on the system performance indicators.

The independent variables/measures representing the communities of practices and e-learning are problem solving, knowledge sharing, and innovation. The dependent variables/outcomes representing the performance indicators are cost

Table 7.4 Contribution of communities of practices and e-learning to improvements in system performance indicators

Dependent outcome	Independent measure that has significant partial regression coefficients	B	Std. error	Beta	*t*	*p*	*Adj* R^2; *p*
Cost competitiveness	Problem solving	−0.38	0.113	−0.393	−2.19	0.034	0.000
	Knowledge sharing	0.455	0.179	0.417	2.545	0.012	
	Innovation	0.335	0.161	0.345	2.075	0.039	
Service and quality excellence	Problem solving	1.721	0.306		5.619	0.000	0.509
	Knowledge sharing	0.607	0.173	0.604	3.510	0.001	0.000
	Innovation	0.221	0.173	0.222	2.798	0.006	

Table 7.5 Contribution of mentoring to improvements in marketing performance indicators

Dependent outcome	Independent measure that has significant partial regression coefficients	B	Std. error	Beta	*t*	*p*	*Adj* R^2; *p*
Profit margins	One to one caring	0.880	0.234		3.762	0.000	0.656
	Trust	0.656	0.066	0.672	9.861	0.000	0.000

competitiveness and service and quality excellence. The results of the analysis are presented in Table 7.4, which shows that the tested regression models are significant ($p<0.05$) and hence the communities of practices and e-learning significantly contribute to the improvement of the system performance indicators.

The results indicate the support for H3. The contribution was significant for all five dependent outcomes with $p<0.05$.

H4. There is significant impact of mentoring on marketing performance indicators.

The measures that represent the mentoring are explained in H2. The dependent variables/outcomes that represent the marketing performance indicators can be regarded as profit margins. Table 7.5 shows the results of the analysis, which supports the hypothesis stating that mentoring significantly contributes to the improvement of the marketing performance indicators as the tested regression models are found to be significant ($p<0.05$).

The results indicate the support for H4. The contribution is found to be significant for all the four dependent variables ($p<0.05$); they for the alternate hypotheses are accepted rejecting the null.

7.6 Managerial Implications

The findings and results of this empirical study are critical for *KM* organizations. The manufacturing sector specifically is faced with growing competitive pressures in a highly disjointed, high market growth and challenges that intimidate its practicality as confirmed by recent consolidations. Effective, continuing, and flexible intra- and interfirm relationships can and should play an important role in an organization's competitive posture. The organization can react to the needs in a timely fashion and also accustomed to the need of market in more efficient manner than its competitors. This study has examined the significant influence and the extent of influence of various learning factors on *KM* performance indicators. It can be seen that the organizational learning contributes largely to the process performance indicators measures. The financial performance indicators are found to be significantly influenced by the expert management system. It is also observed that by focusing these communities of practices and e-learning, the *KM* executives can expect to enhance the marketing performance indicators.

7.7 Summary

Learning designs and models developed in this research have important theoretical contributions to the *KM* literature. In addition to this theoretical contribution, this research also provides important implications for *KM* managers and practitioners, and this design/model can be leveraged as a strategic base for any manufacturing organization in its endeavor to implement *KM* solution. The devised model can be fine-tuned to suit to the needs of the specific goals and objectives of *KM* solution of manufacturing organization. The limitations of the research can be twofold: the learning factors are not designed with respect to human resource and that the base key performance indicators and the learning design can fit only to manufacturing industry. The future scope of this research include exploration and extrapolation of learning design and models for the application of service industry and investigation of other modules such as top management influence design, risk design, training design, and model and sustenance design for implementation of *KM* solution.

Chapter 8
Balanced Scorecard Framework for Knowledge Management Solution Implementation

8.1 Introduction

Planning is very much essential for any project. In the dynamic nature and competitive business world, a measurement tool is very much essential to measure the performance. This is a measurement system in which from individual to organization can assess their performance. The entire task must be linked with vision, mission, and objective of the organization. This will make the organization a strategy-focused organization. Financial measures are inadequate for guiding and evaluating organizations' trajectories through competitive environments. They are lagging indicators that fail to capture much of the value that has been created or destroyed by managers' actions in the most recent accounting period. The financial measures tell some, but not all, of the story about past actions, and they fail to provide adequate guidance for the actions to be taken today and the day after to create future financial value. That is why organizations look toward balanced scorecard (BSC) to sustain growth and to have a harmonious working environment.

8.1.1 Objective for Strategy Planning

Today, organizations are competing in complex environments so that an accurate understanding of their goals and the methods for attaining those goals is vital. Companies are in the midst of a revolutionary transformation. Industrial age competition is shifting to information age competition. During the industrial age, from 1850 to about 1975, companies succeeded by how well this could capture benefits from economies of scale and scope. Technology mattered, but, ultimately, success accrued to companies that could embed the new technology into physical assets that offered efficient mass production of standard products. The information age environment for both manufacturing and service organizations require new capabilities for competitive success. The ability of a company to mobilize and exploit its

K. Ganesh et al., *Design and Development of Knowledge Management for Manufacturing: Framework, Solution and Strategy*, DOI 10.1007/978-3-319-02892-7_8,

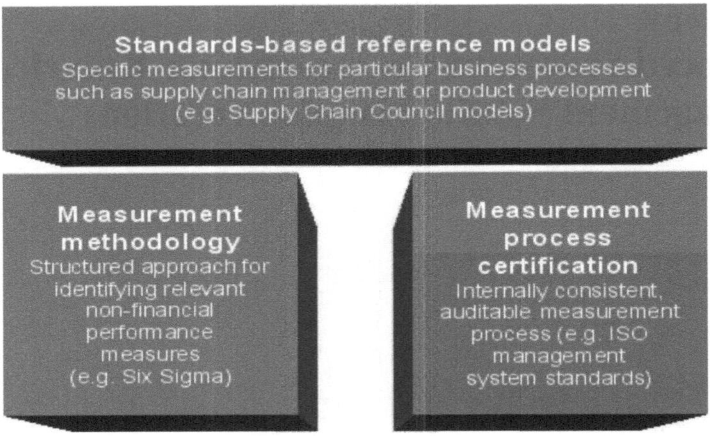

Fig. 8.1 Methods of strategy planning

intangible or invisible assets has become far more decisive than investing and managing physical, tangible assets. So, the BSC framework is essential for the implementation of *KM* solution. These information age organizations are built on new set of operating assumptions like cross functions, links to customers and suppliers, customer segmentation, global scale, innovation, and knowledge workers.

8.1.2 Methods of Strategy Planning

Methods of strategy planning are detailed in Fig. 8.1.

Total Quality Management (TQM), Six Sigma, and EFQM (technically "the EFQM Excellence Model") are among the methodologies and management systems that focus on nonfinancial measurements, which can stand as leading indicators of financial performance. Many case studies demonstrate impressive results following implementations of these systems. A second approach, taken by the ISO and the big public accounting firms, focuses on the internal processes by which the numbers are generated.

8.1.3 Team for Strategy Planning

The team for strategy planning is detailed in Fig. 8.2.

The chief executive officer (CEO) of the company is the one who will frame the strategy. The administrator makes the planning against the strategy and administrates the progress and report to the CEO. Supervisor makes the specification and working progress to the next level. Operation in charge is the person who will carry over the project. The actual working will be done in this area.

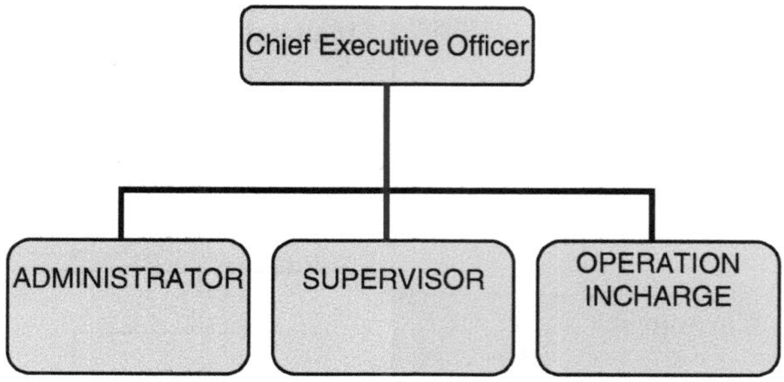

Fig. 8.2 Team for strategy planning

8.1.4 Need for Tool to Assess the Planning

The critical points are:

1. Focusing the whole organization on the few key things needed to create breakthrough performance.
2. Helping to integrate various corporate programs, such as quality, reengineering, and customer service initiatives.
3. Breaking down strategic measures to local levels so that unit managers, operators, and employees can see what's required at their level to roll into excellent performance overall.
4. Fixing stretched target for the measures created and monitoring periodically to improve the results.
5. Attaching these measures to individual performance measure.

The BSC is a tool for measuring a company's activities in terms of its vision and strategies. Fundamentally, the scorecard is a management tool, which continuously reveals whether a company and its employees achieve the results set forth by the strategy. So, the development of BSC framework is critical for the implementation of *KM* solution. But it is also a tool that helps the company express the necessary objectives and initiatives to support the strategies.

8.1.5 Meaning of BSC

"BSC provides executives with a comprehensive *framework* that translates a company's *vision* and *strategy* into a coherent set of *performance measures*," by Drs. Robert Kaplan (Harvard Business School) and David Norton. Recognizing some of the weaknesses and vagueness of previous management approaches, the BSC approach provides a clear prescription as to what companies should measure in order to "balance" the perspective. The BSC is a management system (not only a measurement system) that enables organizations to clarify their vision and strategy

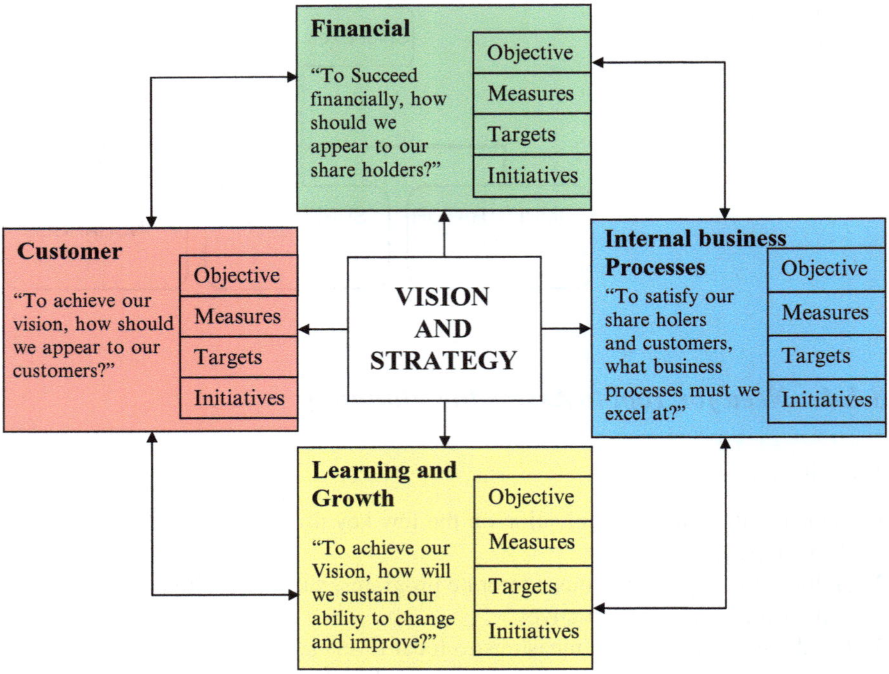

Fig. 8.3 Four perspectives of BSC

and translate them into action. It provides feedback around both the internal business processes and external outcomes in order to continuously improve strategic performance and results. When fully deployed, the BSC transforms strategic planning from an academic exercise into the nerve center of an enterprise. The BSC suggests to view the organization from four perspectives and to develop metrics, collect data, and analyze it relative to each of these perspectives as in Fig. 8.3. The four perspectives are:

- The learning and growth perspective
- The business process perspective
- The customer perspective
- The financial perspective

The goal of making measurements is to permit managers to see their company more clearly – from many perspectives – and hence to make wiser long-term decisions. The Baldridge Criteria et al. (1997) booklet reiterates this concept of fact-based management: "Modern businesses depend upon measurement and analysis of performance. Measurements must derive from the company's strategy and provide critical data and information about key processes, outputs and results. Data and information needed for performance measurement (PM) and improvement are of many types, including: customer, product and service performance, operations, market, competitive comparisons, supplier, employee-related, and cost and financial

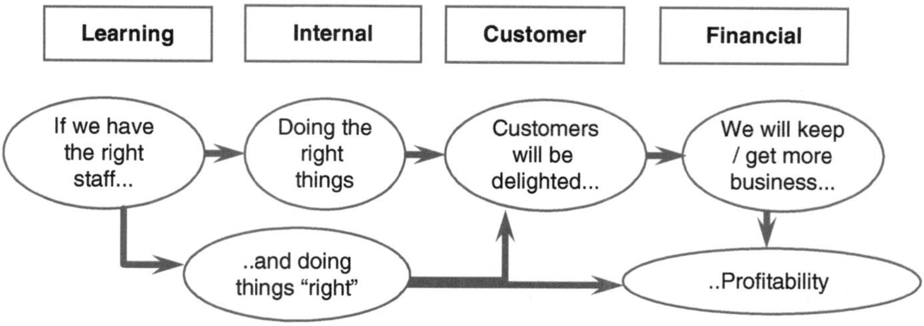

Fig. 8.4 Cause and effect linkages: the basic principle

aspects. Analysis entails using data to determine trends, projections, and cause and effect – that might not be evident without analysis. Data and analysis support a variety of company purposes, such as planning, reviewing company performance, improving operations, and comparing company performance with competitors' or with 'best practices' benchmarks."

A major consideration in performance improvement involves the creation and use of performance measures or indicators. Performance measures or indicators are measurable characteristics of products, services, processes, and operations the company uses to track and improve performance. The measures or indicators should be selected to best represent the factors that lead to improved customer, operational, and financial performance. A comprehensive set of measures or indicators tied to customer and/or company performance requirements represents a clear basis for aligning all activities with the company's goals. Through the analysis of data from the tracking processes, the measures or indicators themselves may be evaluated and changed to support such goals. The basic principle of cause and effect linkage is indicated in the Fig. 8.4.

8.1.6 BSC Concepts

This is a tool to view and analyze the organization's strategic information, including BSC by organization or function, key performance indicators, related cause and effect relationships, and associated graphic displays. The BSC concepts and phases of BSC are shown in Table 8.1 and Fig. 8.5, respectively.

8.1.7 Architecture of the BSC

Oracle BSC uses three-tier architecture which is comprised of a database tier, a web server tier, and a desktop tier. Users connect to BSC through any Java-enabled web

Table 8.1 BSC concepts

BSC framework	BSC management	Organizational feedback and learning
Main views representing strategy objective and perspectives	Key performance indicators	Strategic resources
Cascading multiple scorecards associated with responsibility	Historical trends, comparison, and contribution analysis for multiple organizational dimensions	Strategic communication tools
Cause and effect relationships	Performance measurement at KPI and scorecard level, with alarm colors representing actual and benchmark data	Qualitative assessment mechanisms

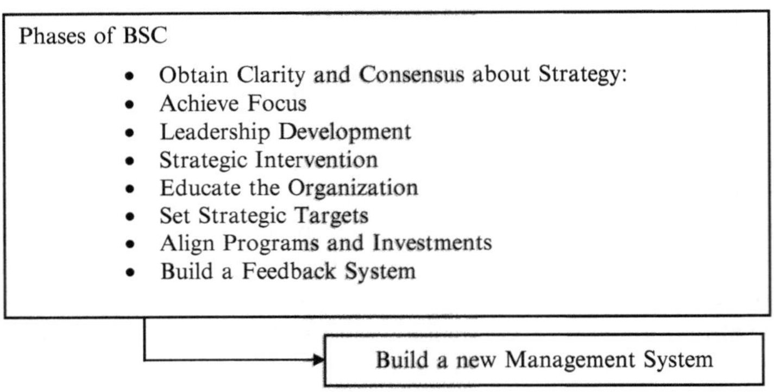

Fig. 8.5 Phases of BSC

browser. The web server tier provides business logic and generates dynamic web pages. The presentation layer of the web server, which is part of the application tier, and the business logic for BSC are implemented as Java Servlets and Java Server Pages (JSP) running on Apache JServ. Apache JServ, a module of Apache web server, implements the Java Servlet API for running server-side Java. The architecture of the BSC is shown in Fig. 8.6.

8.1.8 BSC Architecture Modules

BSC architecture modules are installed in client–server mode connected to an enterprise database. The BSC architecture modules include:

1. BSC Setup: this is the tool used to set up the BSC system after installation. It is used to register BSC systems, administer database privileges, import data, upgrade systems from previous releases, and migrate data between systems.

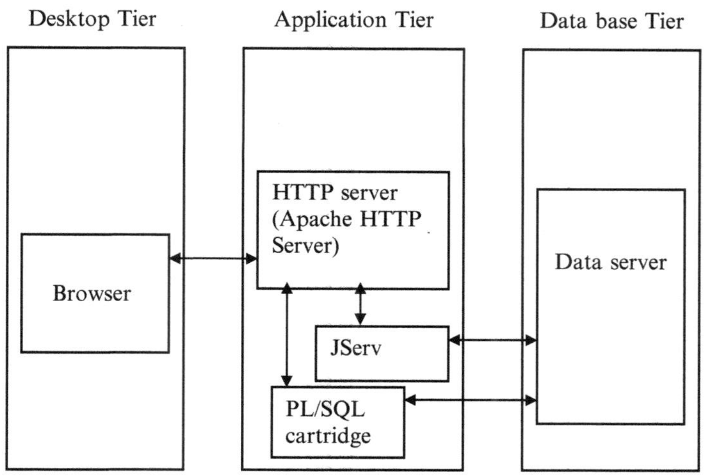

Fig. 8.6 Architecture of BSC

2. BSC Architect: this tool is used to set up BSC; create and configure indicators, groups and datasets; and set calculation and other options inside BSC tables where data is stored. There are three sub-modules:
    - Builder: used to create indicators and groups and to assign key performance indicators (KPIs) to different BSCs.
    - KPI designer: used to arrange how KPIs are viewed and categorized; configure the functionality of scorecard formulas and datasets; and generally tailor the design of the BSC for the user.
    - Metadata optimizer: used to evaluate all the datasets and dimensions specified for a BSC KPI and then create an input table structure for populating data to the BSC. This is run after changes are made in KPI Designer that affects the configuration of different indicators and groups.
3. BSC manager: this tool is used to load and configure the data in the BSC tables. There are two sub-modules:
    - Loader: used to physically load data into the tables used by the viewer from the input tables.
    - Administrator: used to assign and restrict access to BSC (tabs) and KPIs according to responsibility title.
4. System requirements: this section outlines the system requirements for BSC, including requirements for the web browser, software, and BSC architect platform.
5. BSC architect platform requirements: Pentium CPU, or equivalent, with a minimum processor speed of 266 MHz; Windows NT Version 4.0 and Service Pack 3; minimum disk space of 65 MB; minimum recommended memory of 64 MB; BSC architect requires Oracle 8.0.6 Client and Microsoft Excel 95 or above; Windows XP users must install the Oracle 9i Release 2 client and Oracle 8.0.6 client.

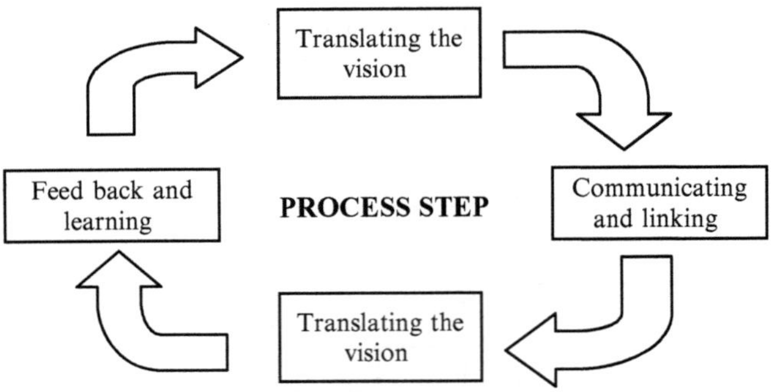

Fig. 8.7 Four processes to implement BSC

8.1.9 Process to Implement the BSC

Using standardized nonfinancial measures will reduce the time, cost, and risk of implementing the BSC. Targets and measures are the "medium of exchange" for each of the four process steps involved in the implementation (Fig. 8.3), and the four processes to implement the BSC are shown in Fig. 8.7.

8.1.10 Need for the Strategy

1. Excessive reliance on a single product (deposits)
2. A cost structure that made it unprofitable to service 80 % of its customers at prevailing interest rates

Strategy against the two needs: Improve operating efficiency by shifting nonprofitable customers to more cost-effective channels of distribution (e.g., electronic banking). Process and development of BSC translated each of these strategies into objectives and measures in the four perspectives. The strategic objectives are shown in Fig. 8.8.

8.2 Research Gap Based on Literature

The need for having a holistic BSC framework for organizational transformation is highlighted in literature. It is also clearly evident that the systematic performance management approach and model through BSC is key for any organizational change like *KM*. From the detailed literature survey, the research gap is shown in Fig. 8.9.

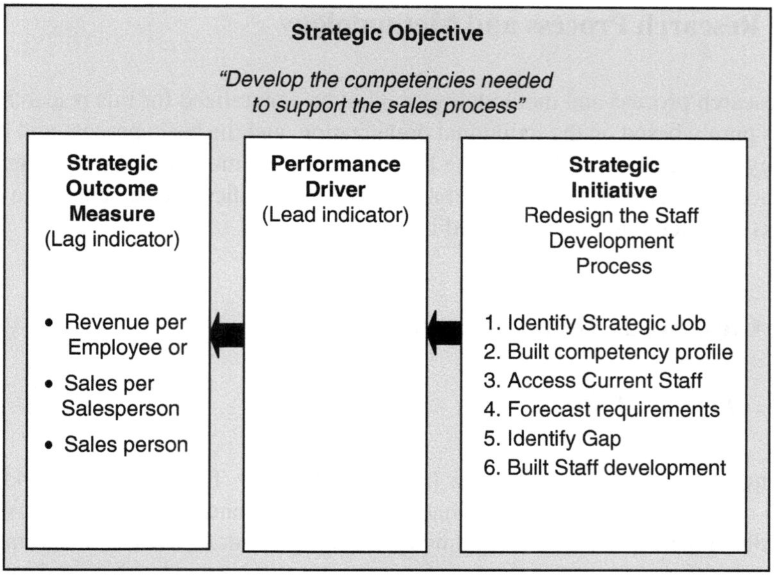

Fig. 8.8 Strategic objective

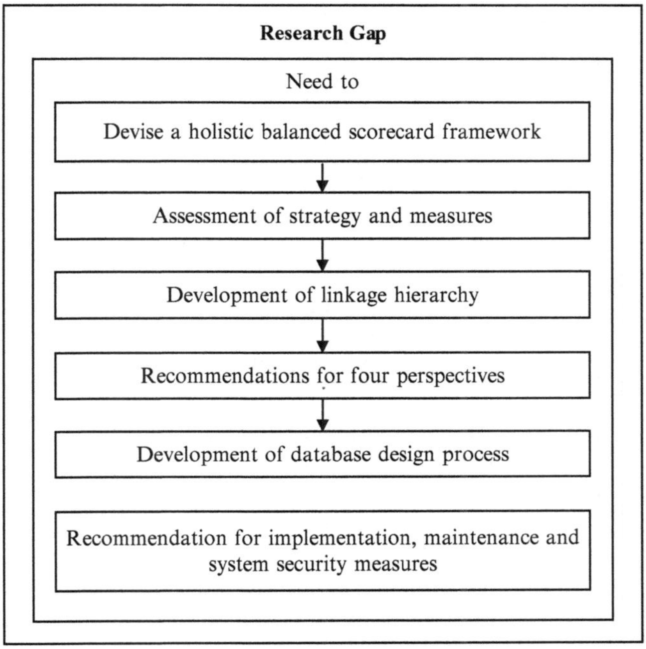

Fig. 8.9 Research gap

8.3 Research Process and Methodology

The research process and methodology cannot be generalized for this research gap. This is purely based on the individual organization, and the basic process and methodology can be developed. From the basic process and methodology, the organization should develop and customize based on their specific requirements. The basic process and methodology is detailed in Fig. 8.10.

8.4 Case Study Demonstration

8.4.1 Period of Study

This study covers a period of 1 year from 2008 to 2009. The past records and base period records are chosen for the trend analysis. Adherence to delivery is based on the budget given by the sales departments. The primary data and secondary data are

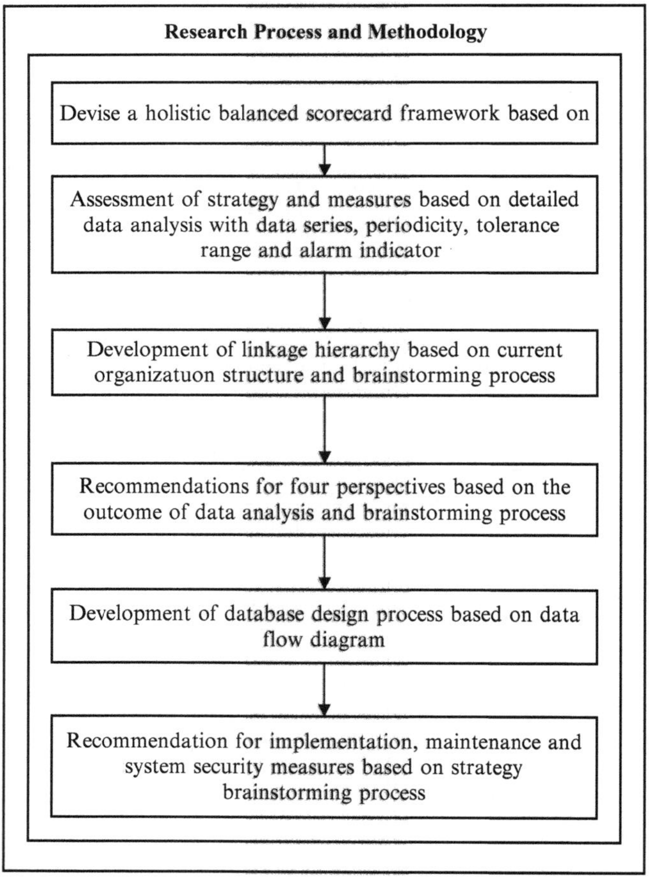

Fig. 8.10 Research process and methodology

used for the study. The primary data were collected from the records directly. The source of secondary is from the journal publications. The data given here is extracted from Electronics Signals and Controls Private Limited.

8.4.2 Data Particulars and Analysis

The data elements include adherence to delivery, customer complaint rate, customer complaint resolution time, revenue from new products, and revenue from new customers. The collected data are classified, rearranged, and regrouped to suit the need for the study. Data sheets are used to collect the data. BSC tool is used for the review of the performance view.

8.4.3 Scope of the Study

Electronics signals and controls (ESICO) is one of the leading manufacturers of various industrial signal lamps catering to industrial needs for handling innumerable applications. Situated in the leading industrial city of Coimbatore, Tamil Nadu (India), ESICO was started in the year 1991 by a team of highly qualified, competent engineers and technocrats from various fields of electronics engineering. Over a span of 13 years with a strong and stable infrastructure consisting of a modernized workshop capable of manufacturing over 55 types of lamps of various ranges and a dedicated work force, ESICO has grown steadily and is currently recognized as one of the leading industrial signal lamp manufacturers. The expertise derived in the field of design, manufacturing, research, and development has been compiled and presented in the form of various products in the domestic market.

8.4.4 ESICO Global Presence

The organization manufactures robust industrial signal lamps catering to various types of industries for intensive lighting in modern machine shops and suitable applications. A strong satisfied clientele of over 200 in domestic market believe strongly on the quality program, strictly adhered to the commitment mission of ESICO. Tight tolerances and insistence on perfection help ESICO to continuously churn out signal lamps. It is an ISO 9000 certified company. ESICO's commitment toward deliverance of superior quality products has been motivated by our core competency of successfully integrating and attuning its research and development standards with superior design and manufacturing capabilities.

8.4.5 Infrastructure

ESICO has been able to consistently produce and deliver world-class products mainly due to its strong infrastructure base. Being a market leader in the manufacture of industrial signal lamps, the organization ensures that our research and development (R&D) facilities are constantly upgraded. The state-of-the-art infrastructure facilities allow us to offer high-quality products. The organization is equipped with a modernized workshop which delivers varied types of lamps for industrial purposes. Excellent in-house training is offered to our work force on a periodic basis. The training imparted greatly enhances the skill sets and enables ESICO to deliver many value-added services to our customers. ESICO is strongly committed to adherence to international quality norms. Sophisticated infrastructure facilities help us to offer quality products at competitive pricing. Our openness to adoption of latest technological standards allows us to attain 100 % customer satisfaction. Our infrastructure competence is matched with the excellent after sales service that we offer. This service network has enabled us to retain and at the same time widen our customer base.

8.4.6 Product Range

The product range is shown in Table 8.2.

This study helps us to commit to performance-based management. Build a strategy for the dynamic situation and progress toward the strategy. "DO right things, DO things Right, DO things Different." This will clearly explain and define each one in the organization a clear role and responsibility toward their work. Ultimately, it provides work satisfaction, and thereby monetary satisfaction as well, and it makes the future organization a "strategy-focused organization."

BSC will bring the following benefits into the organization:

1. Enhance strategic feedback and learning translates a company's strategy into a balanced set of KPIs.
2. Communicate these goals throughout the organization.
3. Align individual, organizational, and cross-departmental initiatives.
4. Enhance strategic feedback and learning.

The BSC for this organization is devised based on the four perspectives.

8.4.7 Adherence to Delivery

The strategy is to reduce cycle time. The measure is based on the budgeted target value versus the actual value for that month. This can be achieved only by satisfying

Table 8.2 Product range

	Conical type LED lamp (L1)
	Miniature conical type LED lamp (L2)
	Dome type LED lamp (L3)
	Reflection type LED lamp (L4)
	Conical type (L6)
	Halogen lamp cylindrical (L16)
	Halogen lamp cylindrical with adjustable base (L17)
	Pyramid lamp (L34)
	Miniature dome type LED lamp (L80)

the customer by knowing the needs of the customer. Customer is the center point of business by concentrating the company's reputation and the stakeholder benefit. This can be achieved by properly monitoring the process from order to cash.

Formula:

Number of machined planned = sum of machines whose scheduled shipment date falls on the given period
Number of machined dispatch = sum of machines whose actual dispatch falls on the given period

1. Data series: value
2. Periodicity: monthly machine dispatched data
3. Tolerance range:

- Unacceptable < 90 %
- Marginal > 90
- Acceptable >= 100 %

Table 8.3 Actual data

Base item	"2005"	"0"	"1"	Delivery (%)	Quantity on time	Total quantity
L100	2005	0	1	100.00	3	3
L600	2005	0	1	90.77	59	65
L600	2005	0	2	71.64	48	67
L600	2005	0	3	64.29	36	56
L600	2005	0	4	66.00	33	50
L600	2005	0	5	78.57	44	56
L600	2005	0	6	90.63	58	64
L600	2005	0	7	76.47	52	68
L600	2005	0	8	79.10	53	67
L600	2005	0	9	92.42	61	66
L600	2005	0	10	86.27	44	51
L600	2005	0	11	85.71	42	49
L600	2005	0	12	92.63	88	95
L100	2005	1	1	100.00	5	5
L100	2005	1	2	100.00	3	3
L100	2005	1	3	100.00	3	3
L100	2005	1	4	100.00	3	3
L100	2005	1	5	100.00	3	3
L100	2005	1	6	100.00	3	3
L100	2005	1	7	100.00	5	5
L100	2005	1	8	100.00	5	5
L100	2005	1	9	100.00	5	5
L100	2005	1	10	100.00	5	5
L100	2005	1	11	100.00	5	5
L100	2005	1	12	100.00	5	5
L600	2005	1	1	100.00	75	75
L600	2005	1	2	100.00	75	75
L600	2005	1	3	100.00	75	75
L600	2005	1	4	100.00	80	80
L600	2005	1	5	100.00	80	80
L600	2005	1	6	100.00	80	80
L600	2005	1	7	100.00	80	80
L600	2005	1	8	100.00	80	80
L600	2005	1	9	100.00	80	80
L600	2005	1	10	100.00	80	80
L600	2005	1	11	100.00	80	80
L600	2005	1	12	100.00	90	90

4. Alarm Indicator:
   - Red – Unacceptable
   - Yellow – Marginal
   - Green – Acceptable

Assessment: Marketing should check the cause and effect for the results, and effective measure should be taken for the improvement. Data sheet for signal lamp and data flow diagram (DFD) symbol are shown in Table 8.3 and Fig. 8.11, respectively.

Fig. 8.11 Symbols in data flow diagram (DFD)

Symbols	**Description**
	Square, defiles a source or destination of system data
	Arrow, identifies dataflow – data in motion
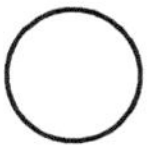	Circle or bubble, represents a process that transforms Incoming data flow(s) into outgoing data flow(s).
	Open rectangle, represents a data store – Data at rest or temporary repository of data

8.4.8 Customer Complaint Rate

The strategy is to proactive management of customer's maintenance schedules and retrofit promotions introduction of annual survive audit with customers, to have more control and supervision on the maintenance at customers end. This KPI gives an indication regarding the number of service request registered versus the number of machines within warranty period. The measure used to monitor is "number of complaints," "number of machines," and "customer complaint rate."

Formula:

- Customer complaint rate = number of service requests divided by the number of machines within warranty

  1. Data series: value
  2. Periodicity: monthly machine dispatched data
  3. Tolerance range:
     - Unacceptable > 100 % with respect to budget
     - Marginal > 90
     - Acceptable <= 90 %
  4. Alarm indicator:
     - Red – Unacceptable
     - Yellow – Marginal
     - Green – Acceptable

Assessment: Everyone's effort must be focused to bring down the customer complaint rate to zero level.

8.4.9 Customer Complaint Resolution Time

The strategy is to minimize resolution time. The measure is based on the budgeted target value versus the actual value for that month. The success of the business is based on after sales service. We must have an effective system to hear the customer complaint either by call center service or online complaint registration. The KPIs give the entire details about the organization's response in resolving the customer complaint against his call. The measure used to monitor is "total number of service request," "time taken to resolve," and "complaint resolution time."

Formula:

- Complaint resolution time = total time taken to resolve the complaint during a specific period divided by the total number of complaints registered

  1. Data series: value
  2. Periodicity: weekly resolution time
  3. Tolerance range:
     - Unacceptable > 100 % with respect to budget
     - Marginal > 90
     - Acceptable <= 90 %
  4. Alarm indicator:
     - Red – Unacceptable
     - Yellow – Marginal
     - Green – Acceptable

  Assessment: Customer service shell should check the cause and effect for the results and effective measure should be taken for the improvement.

8.4.10 Linkage with Corporate Strategy

All the strategies are connected with the key indicators, so that the benefits and the effects can be seen with drill down facility.

Steps involved in link:

1. Main corporate objectives definition with main code.
2. Sub-objectives defined for the departments and divisions align for the main objectives.
3. The data are taken in such a way to achieve the objectives.
4. The main objectives are shown with drill down facility.
5. All the KPIs are concentrated with four perspectives.

The linkage hierarchy is derived based on brainstorming process with 120 executives in the organization, and it is shown in Fig. 8.12

The CEO defines the corporate strategy. The items were distributed across the four perspectives of the BSC as:

1. Customer
   - Adherence to delivery, customer complaint rate, customer complaint resolution time, revenue from new customers, and revenue from new products

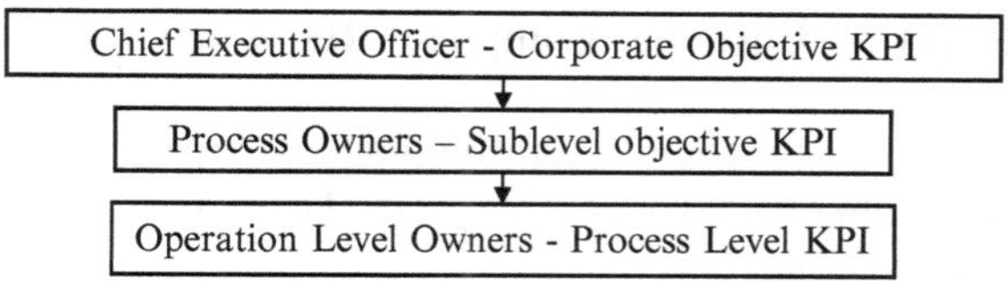

Fig. 8.12 Linkage hierarchy

2. Financial

   - Increase sales revenue, increase net margin; maintain overall margins, asset turnover, return on capital employed; and perform a sales export analysis

3. Internal business process

   - Cost of quality, optimize human capital, reduce cost and inventory control system and manufacturing lean time, learning and growth, training hours per employee, and value added per employee

This will be grouped to the corporate objectives and attached to the individual scorecard. Performance will be updated each month. This will again link with the master scorecard where the CEO can see the achievement of that period, and he can also make comparison with the budget value.

8.4.11 Findings of the Case Study

This study titled strategy planning covers a period of 1 year. The primary and secondary data were collected for the study. The primary data were collected from the records of the mill and the secondary data were collected from the trade journals. The BSC viewer will give the true positions, and the assessment sheet will give what next and how to proceed further.

8.4.11.1 Adherence to Delivery

From the 1-year data analysis, we have noticed the adherence to delivery (delivery of products as per the sales schedule shipment date) was not maintained. We could see the machines scheduled were not delivered; instead, some other unplanned machines were delivered to account for the production volume.

8.4.11.2 Customer Complaint Rate

From the 1-year data analysis, we have noticed the customer complaint rate was more within the warranty period of 6 months to 1 year. This is due to wrong supply and short supply of components. The complaints are occurring because of the urge in attaining the production target.

8.4.11.3 Customer Complaint Resolution Time

The customer complaint resolution time period varies from a minimum of 3 days to a maximum of 30 days. This should be minimized to 6 working days as per management's decision.

8.4.12 Recommendations for the Case Study

The recommendations based on the findings are derived based on brainstorming process with 120 executives in the organization. The above finding initiates four general areas of strategic management activities as follows:

1. Learning and growth is fostered by *KM* activities and initiatives. These include strategic recruiting, hiring, training (both formal and informal), team development, document management, collaborative communication systems, knowledge and skills audits of employees, knowledge base developments, and fostering of communities of interest within the organization.
2. Business process improvements may range from moderate and localized changes to wide-scale changes in business processes, the elimination of paperwork and steps in processes, and the introduction of automation and improved technology. Deployment of the BSC measurement system itself is one of these processes.
3. Customer loyalty cannot any longer be taken for granted within the government nor is it sufficient to manage it in an ad hoc or anecdotal way. Rather, customer relationships are becoming increasingly structured and measured. Not only must the agency work closely with customers on a personal level, it must also gain documented and continuous feedback on customer perceptions and loyalty. These efforts come under the general heading of customer relationship management (CRM).
4. Financial management – in the passive sense of "bean counting" – is giving way to proactive initiatives in activity-based costing (ABC), functional economic analysis (FEA), earned-value management (EVM), and other practices by which managers can learn more from financial data, in order to track projects more closely and make better cost estimates. Also, innovations in budgeting – including the Government Performance and Results Acts' (GPRA) goal of linking performance to budgets – are replacing zero-based budgeting and other earlier techniques in government agencies. The availability of improved database technology with more business intelligence capability is turning financial management into an active part of an agency's overall strategy for success.
5. In conclusion, management experts agree that learning and growth are the key to strategic success, the foundation for the future. Learning and growing organization is one in which *KM* activities are deployed and expanding in order to leverage the creativity of all the people in the organization. The linkage between cause and strategic activities is shown in Fig. 8.13.

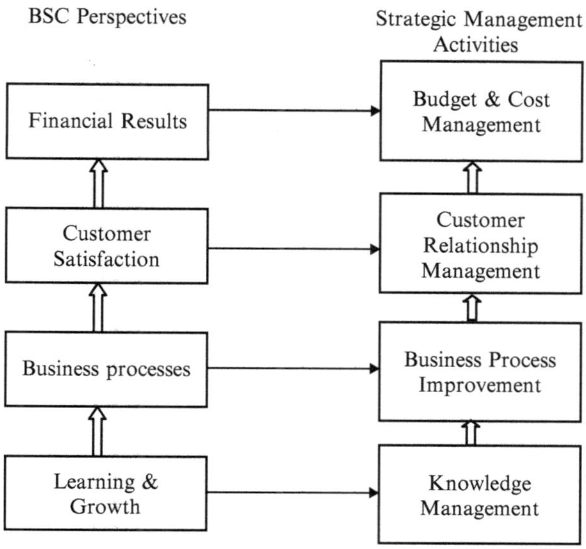

Fig. 8.13 Linkage between causes and strategic activities

Fig. 8.14 Strategy planning

8.4.13 Enhance Strategic Feedback and Learning

Finding of the gap analysis which will analyze the comparison of the desired performance goals with current levels established the performance gap that strategic initiatives can be designed to manage it. Thus, the BSC not only measures change; it fosters change. The strategy planning is shown in Fig. 8.14.

8.4.14 Fundamental Design Concepts

Data flow diagram (DFD) is a charting tool which traces a network of data flows through a system by symbol representation like rectangle arrows, circle, and double

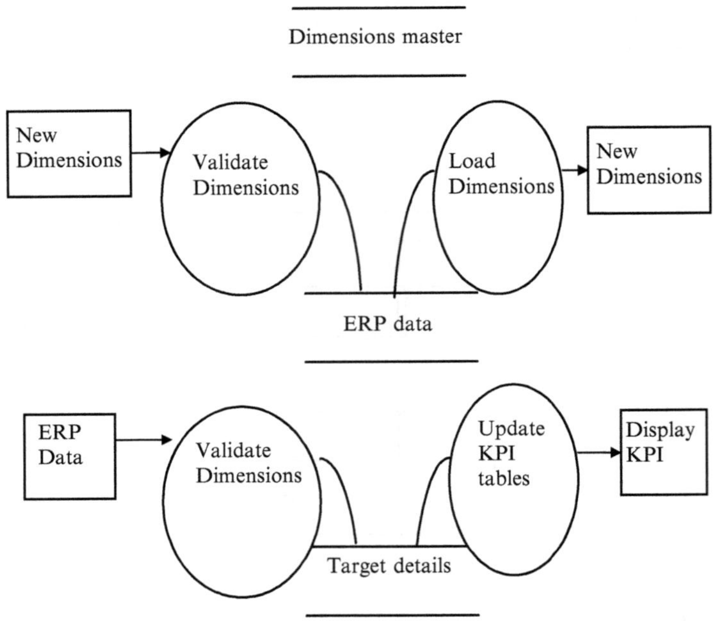

Fig. 8.15 Data flow diagrams

lines. DFD is a documentation technique, which is primarily used in the requirement analysis stage. A DFD can be used at any level of detail. DFD are quite valuable for establishing naming conventions and names of system components such as subsystem, files, and links. It describes what flow (logical) rather than how they are processed, so it does not depend on hardware, software and data structure, or file organization. DFD consists of a series of bubbles joined by lines. The bubbles represent data transformation, and the line represents data flow in the system. It is also known as "bubble chart."

8.4.14.1 Objective of Data Flow Diagram

Easy understanding of the present flow of the business for which the system is built for and DFD reveals any neglected loopholes in the present business operations and presents a clear picture of the user requirements. The DFD is detailed in Fig. 8.15.

8.4.15 Database Design Process

A database is a collection of interrelated data stored with minimum redundancy to serve many quickly and efficiently. The database serves as a repository of data; as

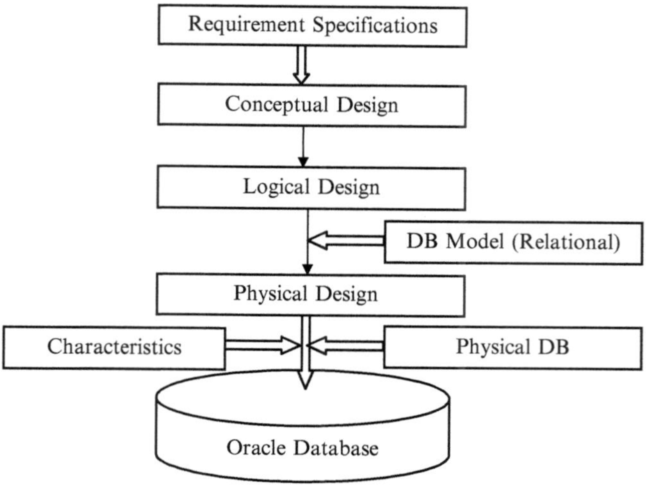

Fig. 8.16 Database design

well designed database environment, common data are available and used by several users. The design of a database is one integrated step in a whole process of system design. The process of database design is splitted into three phases. Take all the requirement specifications that have come and create a design from end users and create a design that is independent of any hardware or software concern. The entity relationship model is most popular. Map the conceptual model into a logical model suitable for RDBMS used. This is a final phase of creating a database. Implement the physical data tables and make decision to enhance the usefulness of the database with the proposed application. The database design is shown in Fig. 8.16.

The summary of BSC design is indicated in Fig. 8.17.

8.4.16 Implementation and Maintenance

The implementation and maintenance points are derived based on the organization strategy and also based on brainstorming process with 120 executives in the organization. The testing process focuses on the logical internals of the software, assuring that all statements have been tested, and on the functional externals conducting tests to uncover errors. This process also ensures that defined input will produce actual results that agree with required results. Various testing strategies adopted in testing the system are explained in this chapter. The strategies for testing include unit testing, integration testing, and system testing. In unit testing, all the program units that make up the system are tested. Unit testing focuses first on the modules, independent of one another to locate errors. This enables to detect errors in coding and logic within the module alone. In unit testing, control paths are tested to uncover errors within the boundary of the module. This testing is also used to ensure the integrity

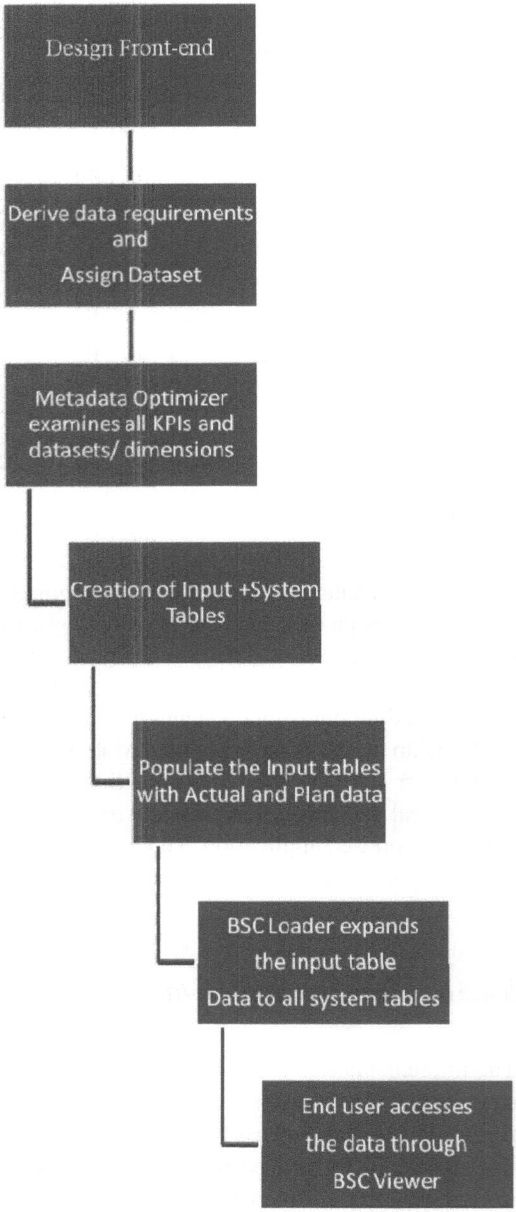

Fig. 8.17 Summary of BSC design

of data stored temporarily. Some of the various test cases used to test the system are as follows: giving inconsistent data for the base table items in the module level and raising unhanded exception cases explicitly and underflow and overflow. Integration testing is a systematic technique for constructing the program structure while at the same time conducting test to uncover errors associated with interfacing. Incremental sandwich integration is adopted in integration testing. That is, the program is constructed and tested in small segments, which makes it easier to isolate and correct. The sandwich approach combines the top-down strategy for the upper levels of the program structure coupled with a bottom-up strategy for the subordinate levels. The interfacing of the software was tested by taking compatible sample data. The interfaces were tested and measures to reduce the response time were taken. System testing is actually a series of different tests, whose primary purpose is to fully exercise the computer-based system. Although each test has a different purpose, all work should verify that all the system elements have been properly integrated and perform allocated functions. The types of system tests are discussed below. Recovery testing is a system test that forces the software to fail in a variety of ways and verifies that recovery is properly performed. If the recovery is done by the system itself, re-initialization, checkpoint mechanisms, and data recovery and restart are evaluated for correctness. Security testing attempts to verify that protection mechanisms are built into the system. Protection mechanisms at OS level, RDBMS level, and application level are tested to avoid improper penetration. This will protect the system from improper penetration and allows authentic personnel only. Stress testing is designed to confront programs with abnormal situations. Stress testing executes a system in a manner that demands resources in abnormal quantity, frequency, or volume. This helps us in fine-tuning the system. Performance testing is designed to test the run-time performance of software, within the context of an integrated system. Run-time performance of the system is tested for all modules, and wherever the performance was poor, alternate simple procedures were adopted. System implementation is the process of making the newly designed system fully operational and not simply installing the software. The following steps are considered in the implementation stage: (1) implementation planning and (2) user training.

1. Implementation planning: This planning is a logical starting point to manage different activities that must be covered. A pre-implementation meeting with the personnel from all departments is arranged. The software is implemented in the month of April 2009.
2. User training: Hands-on training to user is essential to make them comfortable with the system. Accordingly, 2- or 3-day demonstration and practical training with the past data are given to the users who will be the specific users of the system.

8.4.17 System Security Measures

Users are provided with separate user ID and password, and different responsibilities are attached to different users based on the level of access allowed to them. This is the first and primitive way of restricting the user to handle only their data and also effectively ensuring that the user is constraint by not being able to know any other thing other than what is available or shown to them. User login history is stored in a separate file and also in the changed record and can be retrieved on the need basis. This ensures that user logins can be monitored, and also the changes made by them can be located.

In some critical data where more advanced level of security is needed, triggers are built, which writes the changes made in the record to another file with who changed the record with time, old value and new value. This is called audit trail and can be generated on need basis to analyze. Periodically backups are taken and stored in archive and can be recovered on need basis. Different techniques like incremental and full backup are taken based on the strategy designed by the backup and recovery methodology.

8.5 Summary

Plan, set targets, and align strategic initiatives. Senior executives should establish targets for the scorecard measures, 3–5 years out, that, if achieved, will transform the company. To achieve ambitious financial objectives, managers must identify "stretch targets for their internal customer to satisfy the real external customer." Once targets are established managers can align their strategic quality, response time, and reengineering initiatives for achieving the breakthrough objectives. The planning and target setting management process enables the organization to quantify the long-term outcomes it wishes to achieve, identify mechanisms and provide resource for achieving outcomes, and establish short-term milestones for the financial and nonfinancial measures on the scorecard; management experts agree that learning and growth are the key to strategic success, the foundation for the future. Learning and growing organization is one in which *KM* activities are deployed and expanding in order to leverage the creativity of all the people in the organization. The BSC system is an open system. We can incorporate any KPI according to the organizational requirement. The entire system is KPI driven and user friendly, which is useful when worked by all levels of organization. The graphical user interface (GUI)-based output is elegant and easily visualized by all levels. Any system may also have its own drawbacks and can be modified further to incorporate the required changes.

Chapter 9
Vendor Management Framework for Knowledge Management Solution Implementation

9.1 Introduction

Organizations all around the world have woken up to the fact that in order for them to successfully compete in the international markets and to sustain their competitive advantage, they must strive to imbibe and rely on effective supply chains and networks. The management of the supply chain has thus caused a paradigm shift in the way most organizations function (Brandt 2009). Companies are seen to now focus on their core competencies and rope in external suppliers, distributors, and logistics providers in order to ensure that products are manufactured and delivered as per the demands of their customers (Zammori et al. 2009). Hence, it is imperative that there be a close cooperation between the various members who constitute a part in the supply chain.

Vendor-managed inventory (VMI) is one of the supply chain practices in the industry that helps bridge the gap between members of the supply chain and ensures a deeper integration and collaboration among them (Dresner et al. 2009; Disney and Towill 2003a). VMI, also popularly referred as continuous replenishment or supplier-managed inventory, is a strategy based on the principle that the manufacturer or supplier assumes the responsibility for the management of as well as all decisions regarding the product inventory at the customer by utilizing the demand information obtained from the customer (Zammori et al. 2009; Claassen et al. 2008; Chopra and Meindl 2007; Waller et al. 1999). A flow chart of the processes involved in VMI is detailed in Fig. 9.1 (Sari 2007).

The globalization of the supply chain has spurred the growth of VMI. The concept of VMI was pioneered based on the belief that suppliers would be better equipped to handle the customer's inventory due to their adept knowledge in estimating lead times. Accurate and timely information about the expected demand and inventory levels from the customer enables the suppliers to plan production and delivery, preventing stock-outs, improving the visibility of stocks, and reducing inventory costs (Leung et al. 2009; Croson and Donohue 2005). Implementing a VMI therefore discards one echelon of forecasting demand and ordering, dampening the bullwhip effect and eliminating disruptions in the supply chain decision

K. Ganesh et al., *Design and Development of Knowledge Management for Manufacturing: Framework, Solution and Strategy*, DOI 10.1007/978-3-319-02892-7_9,

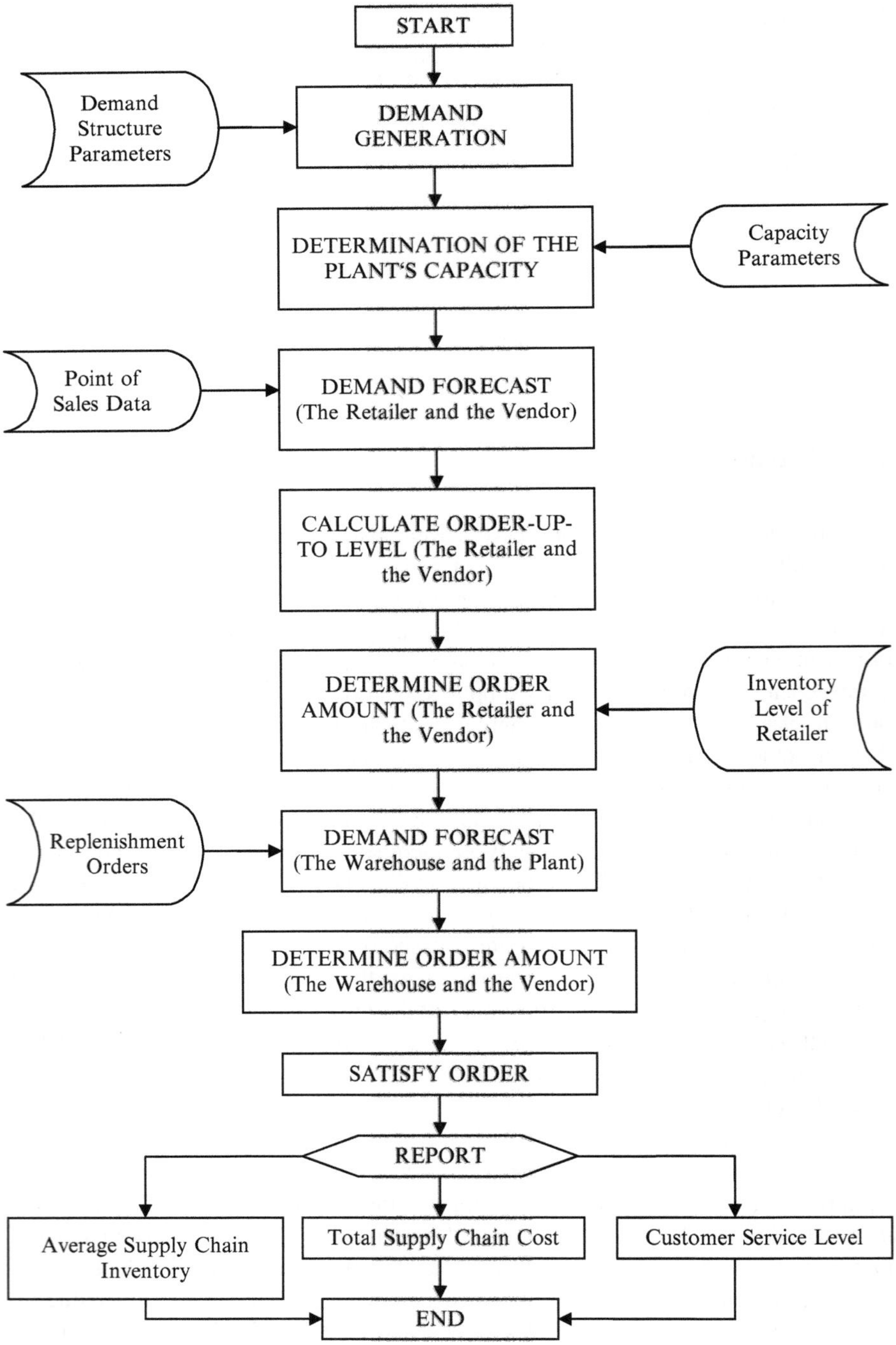

Fig. 9.1 Flow chart of the processes involved in a VMI system

making. VMI is critical for *KM* implementation because the involvement of external entities and external philosophical management approach will improve the effectiveness of the organization.

From the aforementioned flow chart, it can be ascertained that the retailer provides the point of sales (POS) data as well as its level of inventory to the vendor/supplier. This information is processed by the supplier to determine the order amount based on the predetermined order-up-to level and then the supplier proceeds to ship the order to replenish the inventory (Sari 2007). With the removal of one echelon, the inventory cost and total supply chain cost are reduced and the customer service level is enhanced (Michaelraj and Shahabudeen 2009; Belcher 2008).

9.1.1 Benefits for Suppliers

- *Proactive response and alignment of the production* – The VMI system allows the supplier to take over the reins of production planning and stock replenishment for the customer. Information on the demand forecast and existent demand is conveyed to the supplier and this enables them to be proactive and align and plan their production process to replenish stocks (Sari 2007).
- *Reduction in transportation cost* – Implementing VMI enables the suppliers to plan the replenishment schedule and ensure the products are dispatched utilizing the full capacity of vehicles, thereby saving transportation costs (Claassen et al. 2008).
- *Lower inventory costs* – With the clear and precise demand forecast availability, the suppliers are able to reduce their inventory carrying cost as the need for keeping large amounts on inventory as safety stock is eliminated (Saxena 2009).
- *Ease on promotion of newer products* – With suppliers owning the inventory at the customer's premise, it becomes easier for them to introduce new products into the market and avoid the general hesitancy shown by retailers to stock a large amount of newly introduced products, as the demand for them is unknown (Croson and Donohue 2005; Mishra and Raghunathan 2004).
- *Ability to meet changing demands* – With proper demand forecasting, vendors in a vendor-managed system can respond better to changing markets by gauging changing demand patterns (Dong and Xu 2002).
- *Strong and long-lasting relationship across the supply chain* – The suppliers in the VMI system can be ensured to achieve a strong and long-lasting bond with the retailers or end users and reduce the rate of defection from them (Claassen et al. 2008).

9.1.2 Benefits for Retailers

- *Elimination of stock-out issues* – With the supplier taking over the management of the inventory and the timely demand forecasts, the chances of a particular product being out of stock at the retailer's premise are slim (Claassen et al. 2008).

- *Reduction in administration costs* – Implementing a VMI system helps the retailers to avoid voluminous material planning and backorders, thereby reducing administration costs (Sari 2007).
- *Reduction in lead time* – Quite obviously, with the vendor managing the inventory and making replenishment decisions, the lead time for the product to reach the retailers is reduced by a great extent (Wang and Ru 2009; Claassen et al. 2008).
- *Optimum use of space* – Timely replenishment of stocks by the vendor at the retailer's premises prevents buildup of excess inventory, which will take up space in the retailer's premise (Carbone 2009; Waller et al. 1999).
- *Simplified procurement process* – Extensive material requirement planning (MRP) can be avoided by the retailer in a VMI system, as demand forecasts and actual demand are being relayed to the vendor in order to replenish the inventory (Dresner et al. 2009).
- *Reduction of obsolescence* – In a VMI system, the vendor is able to gauge the demand patterns and ensure that the right amount of stocks are replenished, preventing obsolescence of products not sold (Sari 2007).

9.1.3 End User Benefits

- *Improved availability of products* – Using a VMI system ensures that the chances to stock out at the retailers are reduced to a great extent and end users are assured of stock availability when they have a requirement (Xu and Leung 2009).
- *Reduction in switching cost* – Lack of availability of products may decrease and affect the loyalty of a customer, increasing the rate of defection. However, the end user might have to succumb to higher switching costs. This can be avoided in a VMI system.

9.1.4 Benefits for the Entire Supply Chain

- *Minimization of the bullwhip effect* – The ability of the vendors in the VMI system to gauge the actual demand information and demand forecast from the retailer ensures that both the retailers and vendors do not end up with excessive variability in the production and excessive inventory. Thus, in effect, the VMI system plays a pivotal role in the reduction or minimization of the bullwhip, or whiplash, effect (Croson and Donohue 2005).
- *Error reduction* – The robust information systems used for the VMI system help to arrest losses occurring due to human error in data entry (Zammori et al. 2009; Vigtil 2007).
- *Improvement in the processing speed* – The ability of the vendor to receive demand forecasts and the current inventory level on a real-time basis allows them to replenish the inventory without loss of time, thereby considerably improving

the lead times and processing speed in the entire supply chain and (Claassen et al. 2008; Dong and Xu 2002).

- *Lower cost for inventory and transportation at the supply chain level* – As seen earlier, the VMI system helps in the reduction of inventory costs and transportation costs of the vendor. This reduction in costs translates to a reduction in costs at the supply chain level by making the supply chain lean (Zammori et al. 2009; Gentry 2008).

9.1.5 Major Limitations and Risks in a VMI System

In the previous section, the VMI system was seen to have a large number benefits for the supplier, retailer, end users, and the supply chain in general. However, the implementation of a VMI system also poses numerous challenges and risks. Listed below are some of the major limitations and risks of a VMI system.

- *Substitution of products* – Chopra and Meindl (2007) ascertain one of the major limitations in the VMI system is the threat of substitutes of products in the mind of the customer. This was illustrated by taking an example of a detergent manufactured by P&G and a detergent manufactured by Unilever. A retailer implementing VMI for both these suppliers might cause the suppliers to neglect the impact of substitution, which will result in an incorrect inventory level.
- *IT infrastructure cost* – Implementation of a VMI system might require the use of a robust EDI; ERP software such as SAP, i2, and Oracle; or data tracking devices such as RFID and bar code scanners, which are considerably quite expensive to test, set up, and maintain (Vigtil 2007).
- *Management of change* – VMI implementation will bring about major changes in the organization and its processes as the vendor now takes over the inventory management and replenishment. The resulting changes have to be permeated across all the employees in the organization and therefore management of change is a crucial factor in the implementation of VMI (Michaelraj and Shahabudeen 2009)
- *Loss of flexibility* – With the implementation of a VMI, retailers now would be forced to notify suppliers well in advance in case they plan to have special sales events and promotions. This needs to be done so that the suppliers can incorporate the new requirements when deciding on the replenishment schedules. This results in the loss of flexibility for the retailers (Claassen et al. 2008).
- *Threat of forward integration by supplier* – Having the vendor manage the retailer's inventory results in the vendor gaining information of the demand forecasts, actual demands, demand cycles, optimum inventory levels, and sales strategies of the retailer, which results in the looming threat of forward integration by the supplier (Disney and Towill 2003a).
- *Vulnerability to unforeseeable events* – Implementing a VMI system would mean vendors would try and achieve optimum levels of inventory at the retailer's prem-

ise. However, this would address unforeseeable events, such as calamities, requiring more stocks with the retailer (VMI 2009).

- *Dependency and trust issues* – A trustworthy relationship between the supplier and retailer is imperative for the success of a VMI system. A slight hindrance in the relationship can disrupt inventory levels and the entire supply chain (Wang and Ru 2009). Chances are high for retailers to feel that they are being coaxed into holding stocks of the vendors' products only (Carbone 2009). Retailers will also end up with much higher switching costs when they have to switch to other vendors (Brandt 2009).
- *Loss in shelf presence* – Maintaining lower levels of inventory of a particular product, under a VMI system, would at times result in the product going unnoticed by an end user on the shelf, and it is likely that the user might get attracted to products not under the VMI system due to their shelf presence.

9.1.6 Key Issues in Design and Implementation of VMI

A VMI system cannot be designed and implemented overnight. It is imperative for organizations to understand that VMI implementation process needs to be approached in a phased manner. The various phases in the design and implementation of this system are depicted in Fig. 9.2

Phase 1 – Open communication: There should always be an open communication channel between suppliers and retailers on what each entity's strategy, goals, targets, and objectives are when being considered to be part of the VMI system (Zammori et al. 2009).

Phase 2 – Information sharing: As seen earlier, information is one of the key elements in the VMI system, and accuracy of the transmission will play a pivotal role in the success of this system. Suppliers and retailers should therefore reach a consensus to provide accurate information to each other (Vigtil 2007).

Phase 3 – Reliable information receipt: Vendors receive demand forecasts, actual demands, and inventory level updates from the retailers. The use of this information should be primarily to handle the retailer's inventory by the vendor, and therefore the vendors should agree and work on being extremely cautious and reliable with the receipt and use of the information received. Retailers also should ensure that the information transmitted should unarguably be reliable (Vendor Managed Inventory 2009; Vigtil 2007).

Phase 4 – Order policy and risk share: The supplier and retailer should come to a consensus on the minimum order point, maximum level of inventory, and safety stocks. It is also important to reach an agreement on the sharing of risks in the event of stock-outs or obsolescence. Similarly, agreements on the sharing of rewards should be made between the supplier and retailer.

Phase 5 – Resource allocation: The setup and implementation of a VMI system requires a huge commitment of resources from both the suppliers and retailers.

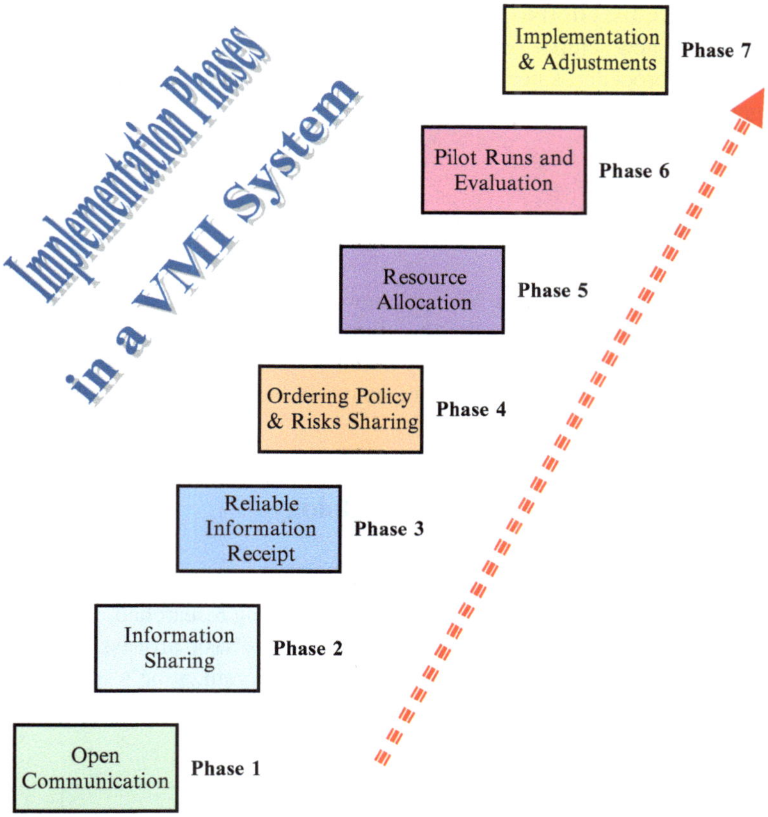

Fig. 9.2 Implementation phases in a VMI system

Personnel from both the retailers and vendors should use their skills, talents, abilities, and knowledge and work together using a robust IT infrastructure to design the VMI system (Ramrakhyani 2009; Vigtil 2007).

Phase 6 – Pilot runs and adjustments: Before the execution and implementation of a full-fledged VMI system, ample pilot runs should be conducted to iron out the flaws and make necessary changes and adjustments in the system.

Phase 7 – Implementation: The final stage in the VMI system is deployment and implementation where the inventory monitoring, planning, and management and replenishment decisions will be taken over by the vendor so as to reap all the benefits described in the earlier section for the vendors, retailers, end users, and the total supply chain. However, the system should be open for continuous improvements and adjustments so that the vendors and retailers can be ready to tackle the ever-changing markets (Zhao et al. 2009; Chen et al. 2009).

Apart from the above steps in the design and implementation of VMI system in an organization, the management should also wake up to the fact that they should put in a considerable effort to help employees adapt to the changes in their environment when this new setup is brought into practice. Appropriate change management processes and procedures need to be in setup for the benefit of the employees and the organization.

9.2 Research Gap Based on Literature

The need for having a holistic VMI framework and evaluation for organizational transformation is highlighted in the literature. It is also clearly evident that the systematic VMI approach and model is key for any organizational change like *KM*. The research gap from the detailed literature survey is shown in Fig. 9.3.

9.3 Research Process and Methodology

The research process and methodology cannot be generalized for this research gap. It is purely based on the nature of the organization being researched, from which the basic process and methodology can be developed. This basic process and methodology can be developed and customized based on the organization's specific requirements. The basic process and methodology is detailed in Fig. 9.4.

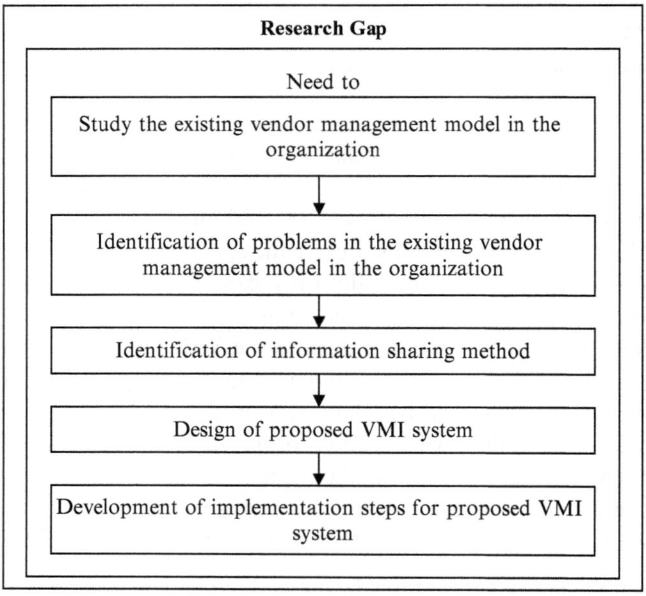

Fig. 9.3 Research gap

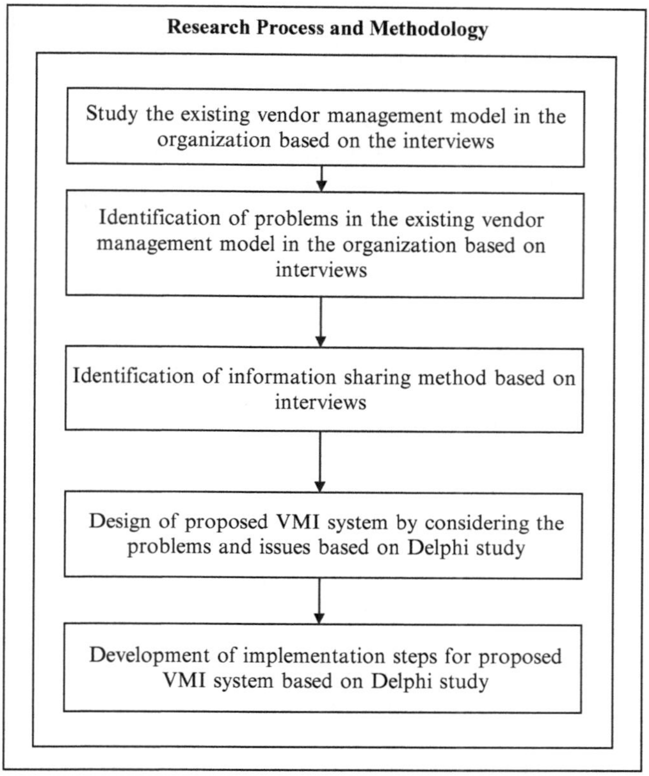

Fig. 9.4 Research process and methodology

9.4 Case Study Demonstration

Concepts like supply chain and vendor-managed inventory are receiving considerable attention today when businesses are struggling globally. Many organizations have realized the enormous benefits that can be gained by introducing these approaches. The concept of supply chain in business promises benefits like reduction in working capital and optimization of resources across various entities of the chain. There are many innovations existing in order to improve the effectiveness of supply chain like vendor-managed inventory, point of sales, collaborative planning forecasting, and replenishment. Vendor-managed inventory became a popular and a powerful tool to remove cost from the supply chain in its practical applicability. It is a streamlined approach to inventory and order fulfillment and is a system in which a vendor continuously and automatically replenishes a trading partner's inventory. True VMI occurs between a distributor/customer and a supplier/manufacturer, with electronic data interchange (EDI) being the crucial link between the two. For any business to survive in this competitive environment, it is necessary to understand the various processes, and they should be mapped in such a way to design the supply chain as per the requirement. Identifying reasons for the gap between the entities in the channel and designing a system to improve the performance of the channel by reducing the gap since *TRUST* are considered to be the critical ingredients in the successful VMI alliance. The single most important benefit of engaging in a strategic VMI alliance could be the chance for cultivating a strong and lasting relationship between the vendor and the distributor, which in the long run can reap rich rewards for both. The scope of the research with regard to Corporation X is as follows:

- To study the existing vendor-managed inventory system between the company and its supplier
- To identify the level of demand met by the existing system
- To point out the problem that affects the performance of the existing system
- To propose a new model to overcome the problems in the existing system

9.4.1 Corporation A Group

The Corporation A Group, with a turnover of over one billion dollars, is the largest manufacturer of automotive components in India. The group produces auto-electrical, diesel fuel injection systems, braking systems, automotive wheels and axle fasteners, powder metal components, radiator caps, two-wheelers, and computer peripherals. Backed by five service and distribution companies with an extensive network across the country, the group has the largest distribution network for automotive products in India.

9.4.2 Corporation X

Corporation X was set up in 1961 as a joint venture of XX Industries to manufacture automotive electrical systems. The company designs, manufactures, and supplies advanced technology systems, products, and services to the world's automotive, after market, diesel engine, and aerospace industries. The combination of these two well-known groups has resulted in the establishment of a vibrant company, which has had a successful track record of sustained growth over the last three decades. Corporation A is one of India's 20 largest industrial houses with 25 manufacturing companies and a turnover in excess of US$ 1.3 billion. The turnover of Corporation X and its divisions was US$ 233 million during 2003–2004. Incorporating the strengths of XX and the Corporation A Group, Corporation X has emerged as one of the foremost leaders in the automotive industry today. Corporation X reaches out to all segments of the automotive industry such as passenger cars, commercial vehicles, tractors, jeeps, two-wheelers, and off-highway vehicles, as well as for stationary and marine applications. With the automobile industry in India currently undergoing phenomenal changes, Corporation X, with its excellent facilities, is fully equipped to meet the challenges of tomorrow. Corporation X manufactures the most comprehensive range of auto-electrical components in the country – a range which continues to set standards in the industry. The products are designed to meet the demands of vehicle manufacturers both in India and worldwide. With the emission standards in India becoming increasingly stringent, Corporation X has ensured that each of its products is manufactured to meet global standards.

9.4.3 *Quality Assurance*

> Corporation X is committed to achieving ever-increasing levels of customer satisfaction through continuous improvements to the quality of the products and services. It will be the company's endeavor to increase customer trust and confidence in the label "Made in Corporation X."

Quality is no longer an option but a basic requirement in today's world. At Corporation X, quality is inbuilt in every phase of manufacturing. The company's quality assurance measures stand on the foundation of a solid belief – that quality begins and ends with the customer. This commitment forms the backbone of its approach to quality assurance. Corporation X has adopted a prevention-oriented quality policy, though ingrained with the traditional ideas of quality control. Everyone from the highest levels of the organization to the lowest practices quality control both as an individual and as a team. An effective quality control system has resulted in the recognition of the company's outstanding achievements in the various fields. Corporation X was awarded the ISO 9001 certification by BVQI in December 1993. The company reached a further milestone when it recently received a certificate of recognition from BVQI for QS 9000 for Auto Electricals.

9.4.4 *Study of Existing VMI Model*

Corporation X's existing VMI model is not well established and not that extensive. The company is trying to develop a well-coordinated model of working. RASIS is the company-owned software. This software is programmed in such a way that it can calculate the bill of materials (BOM) and is connected with the suppliers. Corporation A is known for its quality. It had received Deming's quality award in 2004. So in order to maintain the quality standard, it has a well-established buying group called Y buying group. This group will decide the raw material prices as per the market situation. Suppliers can procure from the approved raw material supplier at the rate fixed by Y buying group. The logistics service for Corporation X is provided by Z Logistics for almost 90 % of the materials. Their duty is to collect delivery indent (DI) from Corporation X to supplier and deliver material from supplier to Corporation X. In some cases, for the remaining 10 %, suppliers manage their own logistics.

Corporation X has a supplier development cell who will monitor the supplier's way of working and their method of procurement. They will even check whether the supplier is procuring materials from and authorized dealer. They will insist all suppliers to plan for their 100 PPM implementation. For new suppliers, the development cell even creates a layout as per Corporation X's expectations. The study of the existing system is based on the interviews with 70 executives in the organization.

9.4.5 Existing Model at Corporation

The existing model at the corporation is explained in detail:

- Purchase department will send a procurement plan to all suppliers every 3 months.
- Marketing will input the demand requirement in RASIS. This will be done every week.
- RASIS will calculate the BOM and it will send indication to the respective vendors.
- Suppliers will plan for their production for that week with the help of RASIS.
- DI will be sent to all suppliers every morning.
- Material will be received from all the suppliers only as per the DI schedule, irrespective of requirement given through RASIS.
- As soon as the material is received, it will be sent for barcode scanning.
- Finally the material is moved into the factory.

9.4.6 Problems in Existing VMI Model

The problems in the existing VMI model are identified based on the interviews with 70 executives in the organization and are detailed below:

- The 3-month plan given by Corporation X, weekly schedule through RASIS, and DI given through Z Logistics differ a lot. This will create a huge problem at the supplier's end.
- Their requirement has huge distortions and the compensation is not borne by Corporation X.
- Critical situations:

  Case 1: Weekly schedule is less than DI; hence, supplier cannot meet the demand for which they have to pay for shortage of material.

  Case 2: Weekly schedule is more than DI, which makes the supplier carry their inventory.

  Case 3: DI will be asked for a material which is not given in the weekly schedule or sometimes material which is supposed to be given the next day as per the weekly schedule.

- So in all the three cases, suppliers are affected a lot. But from the company's point of view, if there are X components required for producing an item A and if one component is not supplied, i.e., if one supplier has not supplied, the whole production will stop. So in order to avoid the termination of production, the company is forced to look for an alternative product which can use the remaining item or by modifying the remaining item. This becomes the root cause of some of the problems.
- This kind of change in the production plan will create immediate requirement of some materials, for which the supplier has to bear the logistics activity.
- Apart from the above-mentioned points, the survey analysis also shows that the suppliers are not protected as they expect. Suppliers are asked to buy raw materials only from those prescribed by Corporation X. When there is an increase in price of raw material, suppliers are asked to pay the increased amount to raw material supplier, but they are not paid for that price increase from the company.
- Corporation X is not only a customer to its suppliers but also a supplier to some of its suppliers.
- In some cases the company itself fails to supply the raw material to its supplier, making the supplier unable to meet his requirement but the company ask the supplier to for it.
- The survey analysis also depicts that the company is not transparent to its suppliers, which is an important factor for establishing VMI.
- Though suppliers are finding a lot of problems, they are also creating problems like increase in the number of returned products and improper response to the software system.

9.4.7 Key Findings from the Study

The key findings from the study are detailed below:

- From analyzing all the problems in the system from supplier's view and also from company's view, it is clear that the major cause is the lack of information sharing, which results in lack of coordination.
- If we look closely into the problem, in some cases the change in production plan is created by a single supplier; because of one supplier's inefficiency in delivery, Corporation X is forced to change the plan, which creates problem for all other suppliers. The point to note here is that not just one supplier is always inefficient, but this could happen with all suppliers, even a good supplier, so the company is not in a position to reject any supplier because of this problem.
- The analysis clearly points out the problem with the software: when the marketing department inputs the demand, the BOM is calculated and the suppliers are intimated as per the demand requirement, but that demand requirement does not include the inventory information.

- Example: Alternator model SM4. Consider each alternator requires one fan.
  - Marketing Input: 220
  - So software will intimate the fan supplier as 220, but the company already has 60 fans on hand, i.e., as inventory; this detail is not updated. This will put the supplier in trouble.
- The software used by the company, RASIS, is updated weekly; hence, suppliers are not in a position to predict day-to-day activities.
- The responsibilities and ownership of the suppliers are very weak in the present working system.
- This system involves huge paperwork on both the company's and the supplier's side.
- Cost involved in communication is very high in the current system.
- Not only the cost but the time involved is also abruptly high.

9.4.8 Different Methods of Information Sharing

Uncertainties in the supply chain usually results in inventory. So in order to reduce the inventory, it is necessary to identify the uncertainties. The reason for uncertainty is the lack of information sharing between members. To reduce uncertainties, the supply chain member should obtain more information about other members. If the members share information, each of them will have more information about others. Therefore, the whole system's performance will be improved because each member can gain knowledge from information sharing. This cooperation mode for increasing information sharing among supply chain members can be called a supply chain partnership.

A supply chain partnership is a relationship formed between two independent members in supply channels through increased levels of information sharing to achieve specific objectives and benefits in terms of reductions in total costs and inventories. It promises a win-win situation for the members involved. The partnerships are focused on the basis of different methods of information sharing (Fig. 9.5) between two adjacent partners on the chain. The average inventory level in the three methods of a typical case study from the literature is graphically shown in Fig. 9.6.

Method 1 is referred to as "decentralized control." The inventories at different sites of the supply chain are controlled independently. There is neither information sharing nor any ordering coordination between the manufacturer and the supplier. Both the manufacturer and the supplier make their inventory decisions based on their own forecasting. The manufacturer uses the customer demand information and the supplier uses the manufacturer's ordering information. We suppose both of them use the base stock policy as their inventory control policy. The base stock policy is of the type with periodic review procedures, which means an order will be placed to replenish the stock level to S at each time period if the stock level is less than the reorder point s. S is called the order-up-to level.

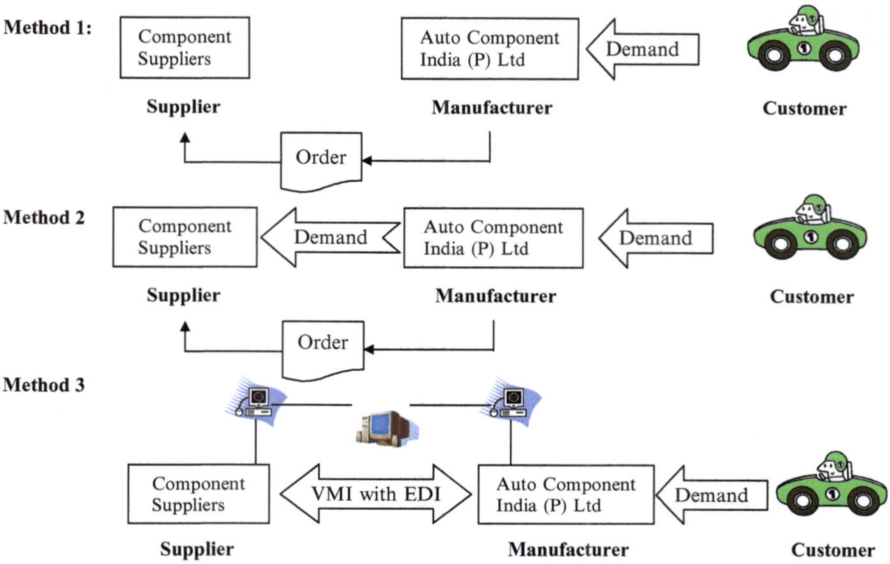

Fig. 9.5 Method of information sharing

Fig. 9.6 Average inventory level in all three methods

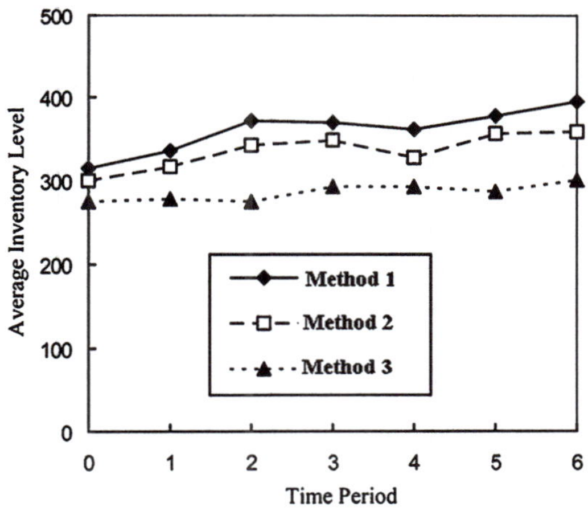

Method 2 is referred to as "coordinated control." The two neighboring inventories are coordinated with sharing of the customer ordering information. In this situation, the supplier will obtain the customer demand information, together with the manufacturer's ordering information, and then make their inventory decision based on both the current customer demand information and the manufacturer's ordering information.

Method 3 is referred to as "centralized control." Here, the decentralized supply chain can obtain the optimal performance achievable by a supply chain under

Fig. 9.7 Five-step process for design of VMI

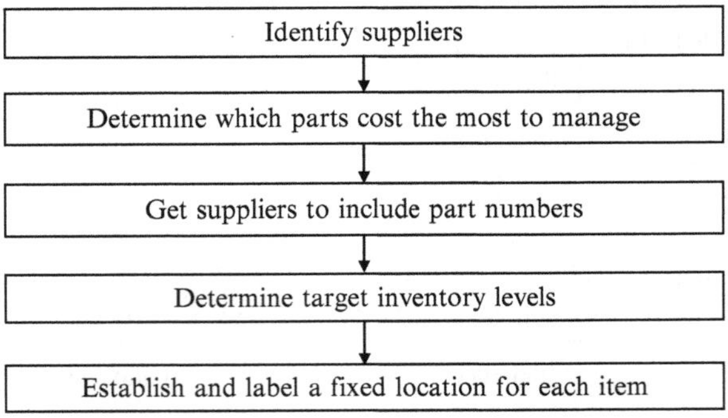

centralized control. Based on EDI, both the manufacturer and the supplier can retrieve the customer's demand information in a synchronized manner. VMI can be adopted. This means the supplier takes the initiative to make major inventory replenishment decisions for the manufacturer in parallel with their own inventory decisions. In this case, the supplier will not depend on the manufacturer's ordering information but on the customer's demand directly.

Pareto improvement implies that all members in a system are at least as well off and some members are better off. From the comparison results of inventory reductions and cost savings of the supply chain members under the three relationship integration levels, it is clear that the supplier can obtain proportionate benefits with an increasing level of information integration. Under the supply chain partnership with information sharing, both the supplier and the manufacturer are at least as well off, and at least one of them – the supplier/manufacturer – is better off. The partnership can improve the overall performance of the supply chain. Pareto improvement can be achieved in respect of the overall performance of the supply chain. Method 3 was chosen for our case study based on the interviews with 70 executives in the organization.

9.4.9 *Design of VMI: A Five-Step Process*

A five-step process is designed based on the Delphi analysis to get its VMI program up and running for direct materials. The flow chart is designed after the third round and is detailed in Fig. 9.7.

The five steps to implement VMI are as follows:

- The *first step* involves identifying suppliers that have the capabilities and willingness to engage in VMI. Selection of suppliers should be based on their proximity to their manufacturing plants, as well as their ability to understand customer's goals and processes, including the macro planning data that customers use to do their own planning internally. (All of the suppliers in the program

are located within an hour of the plant they supply and can get a representative to the plant at least once a week.)
- In the *second step*, the customer should identify the parts from the suppliers that would be good candidates for VMI, focusing first on items with the most frequent deliveries and low value, which cost customer the most. These were items with low risk from a cost standpoint. For example, the process for planning, scheduling, and receiving a component was as long as the process for items that cost only a penny or two. "It is better to look for commodities that are repetitive and are provided by a supplier located nearby, and that could be visually managed."
- The *third step* involves the company and suppliers finalizing the part numbers.
- The *fourth step* is determining order and delivery frequency as well as target inventory levels, including standard Kanban quantities and types of returnable containers that can streamline processes. Kanban container quantities are based on the rate of consumption and the time required replenishing an empty container.
- In the *fifth step*, the company should establish and label a fixed location for each item, as close to the point of consumption as possible, so that the workers would be able to access parts when needed.

9.4.10 Activities Involved in Implementation Process

The supplier representatives should visit the plant at least thrice a week to check bins, slots, or cells and the need for replenishment. The representative visually checks stock levels and checks off material required to replenish the fixed location. The triggering mechanism for reordering can be an empty Kanban container, an empty rack, or an empty location, depending on the part being checked. When items are shipped, they are marked "VMI" by the supplier so the receiving department knows that they are part of the VMI program. To aid in long-term planning, the customer provides VMI suppliers with its macro planning data, e-mailing to them a rolling 12-month demand projection once a week.

For the customer, the benefits of VMI are numerous:

- It eliminates manual planning and scheduling for all material in the program.
- All materials have a fixed and visible location, which the supplier manages, reducing the time to locate material.
- The program eliminates receiving transactions for incoming materials and eliminates the need to maintain delivery information, since the supplier is responsible for proof of delivery information.
- It has also streamlined the company's accounts-payable activities by eliminating the purchase-order/receipt-matching process and consolidating invoices.
- Most importantly, the program eliminates a number of stock-outages.

9.4.11 Benchmarking VMI

VMI programs can reduce inventory by 70 % and increase revenue by 100 %, and the IT investment needed is smaller than you may think. Typical metrics used by the companies to measure effectiveness of the program include:

- Fill rate
- Inventory turns
- Lead times
- Performance
- Demand variability
- Data feed timeliness and accuracy

9.4.12 Supplier Segmentation

If vendor-managed inventory (VMI) is implemented properly, there is potential for a great partnership that will yield immense value for both the customer and the vendor. However, if one party goes into a VMI engagement thinking they will be the sole winner, the probability of success is greatly diminished, as it is a fact that the termination of any type of partnership can be expensive and, to use an all-encompassing term, messy. VMI is a relationship where the customer's inventory levels are monitored and replenished by the vendor, based on a service contract. This must be a contractual relationship set up with the intent of protecting the interests of both the customer and the vendor. The items that need to be defined in the contract include demand forecast and consumption methods and timing, the length of the contract term, prices and annual quantities of products covered by the contract, implementation methods and timing, invoice payment terms, service levels and penalties, termination methodology, quality methodologies, and any engineering change notification/timing methods that need to be employed. But implementing VMI immediately with all suppliers is a difficult task. So it is essential to identify potential suppliers or to segment the suppliers. The following model will illustrate the method of segmenting the suppliers.

9.4.13 Empirical Model of Supplier Segmentation

The model consists of two dimensions that underpin a VMI relationship strategy toward suppliers:

- The supplier's commitment to VMI
- The commodity's importance to VMI

Family* → *Friend* → *Business* → *Transaction

Fig. 9.8 Implementation Step 1 of supplier segmentation

Fig. 9.9 Implementation Step 2 of supplier segmentation

		Supplier's Commitment to VMI	
		High	Low
Commodity's Important to VMI	High	Family	Business partner
	Low	Friendly	Transactio nal

Each dimension is divided into two categories referring to a high or low degree of the supplier's commitment, and a high or low degree of the commodity's importance, to VMI. The generic model of supplier segmentation consists of four principal relationship strategies emphasizing the supplier's commitment, and the commodity's importance, to a VMI. Each relationship strategy has its own characteristics and expectations from a VMI point of view in terms of supplier selection criteria:

- A transactional relationship strategy signifies that the manufacturer invests limited resources in this specific supplier relationship and in VMI. The supplier delivers only single, simple components. This relationship strategy is characterized by low mutual commitment, alternative supplier choices available, and often price-driven transactions.
- A friendly relationship strategy signifies that the manufacturer continues to foster a strong relationship with a specific supplier. The supplier is considered as a partner to the manufacturer. This relationship strategy is characterized by the supplier being dedicated to the VMI and, in fact, often dependent on the manufacturer, while not so innovative.
- A business partner relationship strategy signifies that the manufacturer maintains a high level of competition between one supplier and the others. This supplier is one of the larger ones that deliver to the manufacturer. This relationship strategy is characterized by the fact that the supplier is usually a market leader, significant buying amounts are involved, and the supplier has a range of product offerings.
- A family relationship strategy signifies that the manufacturer invests resources in VMI in one specific supplier relationship and also develops strong corporate partnership with this principal supplier. This relationship strategy is characterized by commitment to mutual success between the manufacturer and the supplier, strategic for technology advancement, critical to the manufacturer cost success, and important to the brand of the manufacturer.

The three steps of flow of implementation are detailed in Figs. 9.8, 9.9, and 9.10.

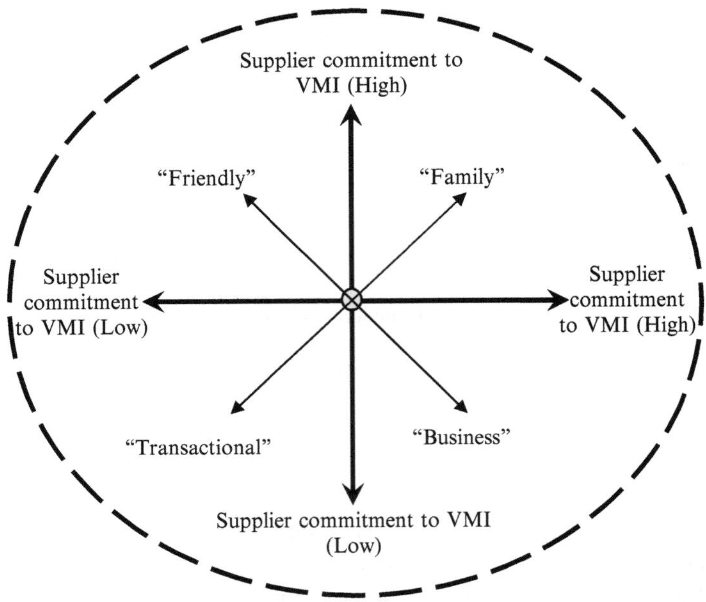

Fig. 9.10 Implementation Step 3 of supplier segmentation

9.4.14 Customer Business Processes

The customer must have some method of communicating demand and consumption with the vendor. This may be a situation where the customer regularly sends a forecast of demand via EDI or B2B Internet communications and then also updates actual consumption and stock adjustments in the same manner. Alternatively, there may simply be an annual contract in place with on-site replenishment taking place on a predetermined basis. This is quite effective in manufacturing environments where standard products are used in multiple floor-stock locations. There can also be a combination of demand/consumption methods and purchase committal dates that are typically defined at the product level. Whatever the situation, the methodology and timing must be documented in the agreement.

In order to manage the contract automatically, the customer should have contract purchase order functionality in their ERP system. This functionality includes the ability to maintain the start and end dates of the contract; the annual quantities expected, and the price per unit of measure for each product contracted; the release mechanism for each product (e.g., forecasted, order point); the agreed-upon service level or reorder point; the stock authorization type (e.g., approval or automatic); the payment terms; cancellation criteria; stock-out penalty info and capture of incidents; and, whether all, any, or none of the products are QAS (Quality-At-Source) certified. QAS is a methodology in which customer and vendor share information that allows the vendor to pre-inspect material based on the customer's specifications prior to shipment.

The purchase-order system must also be able to accept multiple receipts against one order and track the receipts by part number when there are multiple parts covered by the VMI contract. A workaround is to have one contract purchase order for each part number covered. The downside is the number of orders that need to be maintained and the amount of work required to set them up.

The performance of both the vendor and the customer needs to be monitored through reports from the purchasing system. Obviously, vendor performance is measured on the ability to maintain stock levels per contract and the quality of product, etc. However, the purchasing system should be integrated with the accounts-payable system and be able to monitor on-time payment of invoices so that the customer can readily show compliance to invoice payment terms. Additionally, the purchasing system should be able to flag the buyer when consumption of product either exceeds or lags behind forecasted demand levels. The customer should also undertake cycle-counting the product receipts as an audit of compliance.

9.4.15 Vendor Business Processes

Depending upon the type of communication agreed upon for demand forecast and consumption, the vendor must be able to receive and process the information in the same format as the customer. The interesting part is that not all ERP solutions can differentiate demand forecast by customer. However, this is a function that is essential to tracking forecast deviations, planning replenishment, and potentially tracking "commit to buy" aspects of the supply contract. The customer order system must have all of the same elements as the corresponding contract purchase-order system. If the customer order system is not able to track multiple shipments against multiple lines, then multiple sales orders will be raised against a single customer purchase order. The lack of multiline customer order contracts with multiple shipment capabilities will complicate reporting requirements. Customer orders must also be integrated to the quality system should a QAS environment be required. In most QAS environments, a certificate of compliance or certificate of analysis is required with each shipment.

9.4.16 Proposed VMI System and Necessary Steps for Implementation of VMI

The proposed VMI system and the necessary steps for implementation of VMI are derived based on the results of the third round of the Delphi analysis.

Step 1: The top management must commit to the process. Their support is vital. They must commit to factors like costs involved, manpower needed for setup/maintenance, and also the concept of having someone else manage their inventory.

Step 2: The employees should adapt themselves to this new concept. Without their acceptance, the VMI program cannot take off. They must understand that VMI will not push them out of a job. It will free up some of their time to allow them to be more productive in other areas.

Step 3: Synchronize the customer's product files with the supplier. This step alone is one of the greatest benefits of VMI. This is done by establishing new ERP software with all required information. Synchronizing means that you must match the supplier's product data with the distributor product data. Are there old, obsolete items on the file? Have new product numbers been properly communicated to the distributor?

Step 4: Extensive testing of all EDI sets to be used. The supplier and distributor/customer must work very closely together to validate that the data is being properly sent/received. For example: Does the quantity on hand that is being received by the supplier match the quantity on hand in distributor's stock? EDI testing may take many tries and adjustments before its finalization.

Step 5: The distributor/customer must understand and agree with the stocking plan the supplier is creating. Even though the exact method may be a proprietary method, the distributor should still have an understanding of how the plan is created. This will help avoid the future question: "Why did they send us this product if we don't need it?" Additionally, predetermined inventory turns, fill rates, and service levels should be targeted. The distributor/customer should also monitor their current performance for comparison with later results. Both parties must agree upon the frequency of replenishment (once daily/twice a week).

Step 6: The distributor/customer sends the supplier his point of sale history file, usually for 1–2 years (on a disk or through e-mail). The format of the file must be compatible with the needs of the supplier. Then, the distributor sends an EDI indicating the product movement. This tells the status and stock level of every item they have. Both sets of data have to be verified. This is the last and most important validation point.

Step 7: Distributor/customer should use the item and enter that transaction into the computer.

Step 8: On a daily/weekly basis, the distributor/customer sends a product activity.

Step 9: The supplier receives the material movement information and demand requirements and updates the distributors' stocking plan. Once an item or items have hit their reorder point (ROP), the supplier creates an order.

Step 10: The supplier electronically generates a purchase order to the customer. During the beginning stages of the VMI partnership, it is important to have periodic reviews at both ends to identify any problems.

Step 11: The supplier collects and ships the order and transmits a delivery notice. When the shipment is received, the customer transmits a material receipt notice (MRN). This tells the supplier exactly what was received. The supplier can then match this to his purchase order to determine any potential problems (mis-shipped, etc.)

Step 12: The invoice is sent out via an electronic billing system. Payment is done through electronic transfer of fund (ETF).

The flow chart for the proposed system is detailed in Fig. 9.11.

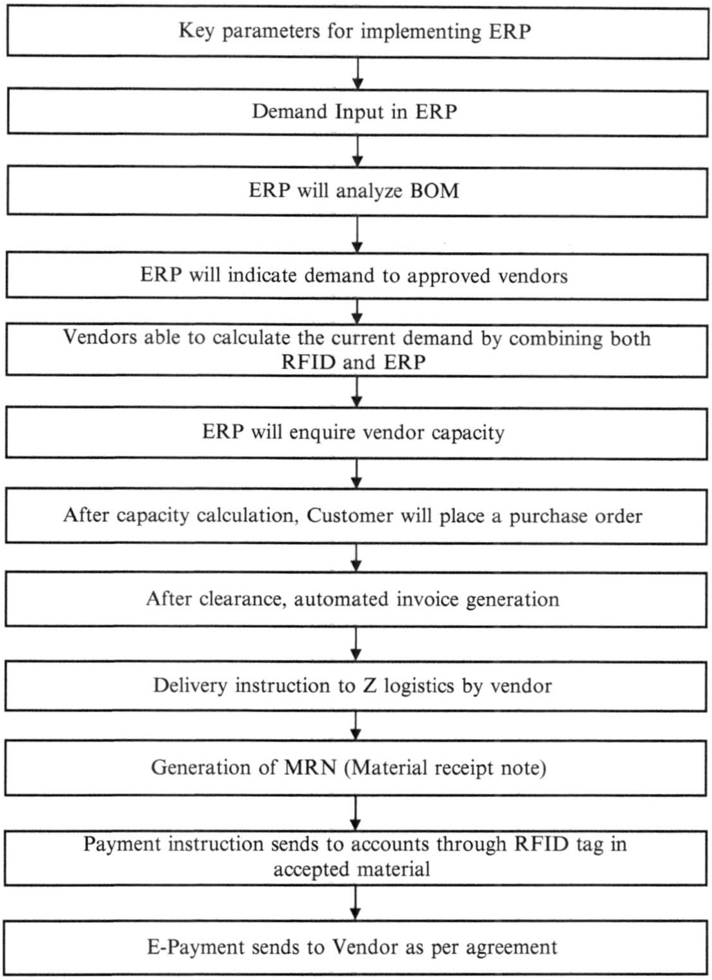

Fig. 9.11 Proposed VMI model

9.5 Conclusion

The importance of managing an organization's supply chain and aligning the supply chain strategy with the competitive strategy of the firm have become pivotal factors in achieving and sustaining a competitive advantage against other organizations in the same industry. The VMI is an apt supply chain practice to analyze and study. The need to attain a lean supply chain, with reduced lead times and lower levels on inventory, is what most firms try to achieve. The VMI system therefore is a key supply chain practice which helps firms to achieve their goals.

The long-term success of a firm depends on the success of its suppliers and the level of satisfaction of its customers. That is, the entire supply chain must be successful. VMI is a part of supply chain. What is being experienced today is a fundamental change in global business philosophy of increasing partnering arrangements. In order to manage these supply chain arrangements for realizing overall improvement in enterprise productivity, it is necessary to improve the planning and management of complex interrelated systems such as materials planning, inventory management, supplier relationship, supplier development, capacity planning, logistics, and production systems. The availability of information technologies has enabled the delivery of integrated systems for decision making.

It is important to establish strategic partnerships with suppliers for a successful supply chain. Corporations have started to limit the number of suppliers they do business with by implementing vendor review programs. These programs strive to find suppliers with operational excellence so the customer can determine which supplier is serving better. The ability to have a closer customer/supplier relationship is very important because these suppliers are easier to work with. With the evolution toward a sole supplier relationship, firms need full disclosure of information. They may establish a comparable culture and also implement compatible forecasting and information technology systems. This is because their suppliers must be able to link electronically into the customer's system to obtain shipping details, production schedules, and any other needed information.

The proposed system integrates the supplier end with the company thereby sharing all the necessary information for the better operation of the channel. Since the proposed system is completely systematic and there is no manual process, there is a better chance of improving the channel efficiency. Though the process seems simple to implement, in the real world of personalities and professional relationships, there are many obstacles to overcome. In short, trust is very important for the VMI model to succeed. Paramount to the success of VMI is the incorporation of technological tools like EDI.

VMI partnership is a relationship formed between two independent members in supply channels through increased levels of information sharing to achieve specific objectives and benefits in terms of reductions in total costs and inventories. It promises a win-win situation for the members involved. Various suggestions are depicted for the proposed system to enhance the performance of VMI. But, the effective enhancement of VMI also depends on the nature of items and materials used in the industry. So, VMI can be implemented effectively only through stepwise improvement. The evolution of present-day market and the change in roles and power within the channel have transformed competition between firms into competition between entire supply chains. Focusing strategy on improving the performances of channel is the only road the firm can take to gain greater competitive advantage. Hence, the proposed system will improve the working nature and performance, but for continuous improvement of the system in the future, it is better to modify the system by taking different factors like nature, cost, and shelf life of the items and materials into consideration.

Future research prospects are detailed below:

- Development of the conceptual VMI model for multistage supply chain with vertical information integration
- Simulation, optimization, and comparison of VMI model for inventory decisions with traditional system
- Simulation of the conceptual VMI model using both periodic and continuous review policies
- Development of a heuristic for optimizing the system parameters of inventory decisions
- Comparison of the traditional inventory model with a VMI-based inventory system

VMI is more likely to lead to higher supply chain profits if both parties commit to share precise internal accounting information and reliably transmitting, receiving, and using this information for inventory decisions (Kulp 2002).

Chapter 10
Conclusions

10.1 Summary

The successful implementation of *KM* solution solely depends on three pillars, namely, people, process, and technology of the organization. There are various elements that influence the people, process, and technology of *KM* implementation solution, namely, readiness, behavior, taxonomy, technology, structure, environment, reward, learning, communication, internal and external levels, and functions of organization. This study was undertaken to address and develop the designs and models required for the above critical elements for the implementation of *KM* solution. A temporal confirmation conceptual model was designed based on the macrolevel brainstorming study, and six modules are identified for the microlevel study. Integrated approach of empirical and expert opinion study was conducted for each module to devise and generate a generic design and process blueprint of identification of readiness level of people, process, and technology for *KM* solution implementation, behavior assessment of people for the change, current and future taxonomy and technology landscape required for *KM* solution, process and environment design along with learning, reward and communication design for the *KM* solution implementation, and linkage design for internal and external levels and functions of organization, namely, balanced scorecard and vendor-managed inventory. Designs and models developed in this research have important theoretical contributions to the *KM* literature. In addition to this theoretical contribution, this research also provides important contributions for *KM* managers and practitioners, and all the designs and models can be leveraged as a base for any manufacturing organization, which is willing to implement *KM* solution. The devised models can be adjusted to suit to the needs of the specific goals and objectives of *KM* solution of manufacturing organization.

The primary intention of readiness assessment is to assess the exact status of where the organization stands today with respect to key critical success dimensions of three pillars of *KM*, namely, process, people, and technology. The objective of this module is to design a generic conceptual framework and generic solution procedure

K. Ganesh et al., *Design and Development of Knowledge Management for Manufacturing: Framework, Solution and Strategy*, DOI 10.1007/978-3-319-02892-7_10,

for readiness assessment for any manufacturing organization with a set of standard components and subcomponents. The components and subcomponents are derived from the literature and those can be changed with respect to mission and vision of the organization. Readiness assessment was carried out for the case organization with 85 executives, and it is suggested that the organization needs to improve the readiness in the areas such as tacit knowledge sharing, explicit knowledge sharing, knowledge centers establishment, and right measures for *KM* solution.

The primary intention of behavior assessment is to assess the exact behavior pattern among the people in the organization toward the implementation of *KM* solution. The objective of this module is to design a generic conceptual framework and generic solution procedure for behavior assessment for any manufacturing organization with a set of behavior types and targets. The types and targets are derived from the literature and those can be changed with respect to mission and vision of the organization. Behavior assessment was carried out for the case organization, and it is suggested that there are six behavior types in the organization, namely, skeptic (10 %), convert (23 %), cynic (14 %), procrastinator (16 %), potential (21 %), and rebel (16 %), and the targets for certain behavior types need to be reviewed and revised for the improvement toward the acceptance for *KM* solution implementation.

The primary intention of designing taxonomy and technology architecture is to identify the current and future *KM* components in navigation and content layers and to access the suitability of technology landscape toward the implementation of *KM* solution. The objective of this module is to design a generic conceptual framework and generic solution procedure for devising taxonomy and technology architecture for any manufacturing organization with a set of *KM* components and IT integration landscape. The *KM* components are derived from the literature and those can be changed with respect to mission and vision of the organization. Taxonomy and technology architecture was designed for case organization with 1 navigation layer, 20 content layers, and 167 *KM* components for structured knowledge. Macrolevel technology architecture was suggested.

The primary intention of process and environment design is to devise the process design for knowledge capture, storage, retrieval, administration, key performance indicator, and organization structure for the implementation of *KM* solution. Process design and organization structure was designed for the case organization.

The primary intention of learning, reward, and communication design is to devise and design the learning factors, incentive plan, and metrics and communication plan for the implementation of *KM* solution. This module was also designed for the case organization.

The primary intention of linkage plan for internal and external functions of organization is to provide a platform to link the *KM* goals to the overall goals of the organization and also to link to external entity of the organization. We have considered one internal function which is the balanced scorecard and one external function which is the vendor-managed inventory. Balanced scorecard is a strategic planning tool which can align the organization to the vision of success and get people working on the right things and focusing on results. Development of balanced scorecard system is like putting a puzzle together, where different pieces come together to form a complete mosaic. We developed a conceptual framework and model for the implementation of balanced scorecard with the focus on customer relationship-based strategic planning, which is a part of *KM* solution implementation. The conceptual framework for a case company was successfully implemented with the focus on architecture for balanced scorecard. Vendor-managed inventory (VMI) is one such supply chain external function where the vendor is given the sole responsibility to monitor and manage the inventory of the retailer. This research attempts to analyze this supply chain external function and illustrate the key knowledge elements of this system and the role it plays in the overall supply chain. The benefits and risks of the VMI system are also analyzed. The design and implementation steps for VMI are extensively reviewed. A case study with a view of *KM* solution implementation is discussed. The analysis of industries which would benefit from implementing this system is also conducted.

10.2 Scope for Further Work

The future scope of research includes exploration of design and models for the application of service industry and investigation of other modules such as top management influence design, risk design, training design, and model and sustenance design for implementation of *KM* solution. It is not clear what kind of organizations will survive in future environment or what types of learning will be central. The increasingly dynamic internal and external environment of organizations results from the growing intensity of knowledge generation in and around organizations. So, the future research for exploration of impact of external components for the implementation of *KM* is critical. The detailed research on *KM* functions is also important. KM is a term that received some negative press because many organizations spent significant amounts of funding on them in the 1990s and they didn't quite work as expected. The future research on understanding the threats for *KM* implementation is critical. This research does not address microblogs as KM systems but the connection is important to point out for potential further research and microblog development. At the intersection of user and system level, most research tends to investigate motivations of users. There is a lack of research on usability of *KM* system and limited studies on usage of *KM* system. Both usability and usage studies if well designed can provide a good indicator of user acceptance of *KM*S. There appears to be a relative paucity of *KM* evaluation studies at the group and team levels except for a few virtual team studies. Although there have been studies at the project level which could be interpreted as group-level evaluations, these studies did not investigate group characteristics and team dynamics in relation to evaluation of *KM*. This area presents an opportunity for future research on team effectiveness in terms of *KM*. There appears to be a lack of studies focusing purposefully on evaluation of *KM* strategy and *KM* structure. Considering that both elements can be vital to the success of *KM* initiatives, research on these elements is required. Aggregation from user- and system-level evaluation to team-, project-, and business unit-level evaluation and subsequently to organization-level *KM* evaluation could provide a worthwhile avenue for future research.

References

Achabal, D.D., Mchintyre, S.H., Smith, S.A., Kalyanam, K.: A decision support system for vendor managed inventory. J. Retail. **76**(4), 430–454 (2000)

Alas, R.: Organizational changes and learning in production and services organization during a transition in society. Int. J. Serv. Oper. Manag. **4**(4), 481–497 (2008)

Alas, R., Vadi, M.: The impact of organizational culture on organizational learning and attitudes concerning change from an institutional perspective. Int. J. Strateg. Chang. Manag. **1**(1/2), 155–170 (2006)

Alas, R., Vadi, M., Sun, W.: Connections between factors of readiness to change in Chinese organizations. Int. J. Chin. Cult. Manag. **1**(2), 219–231 (2008)

Ambjorn, N., Miguel Angel, S., Miltiadis, D.L.: Learning processes and processing learning: from organizational needs to learning designs. J. Knowl. Manag. **12**(6), 5–14 (2008)

Andrew, R., Tom, B., John, C., Andreas, S.: Deep learning design for sustainable innovation within shifting learning landscapes, vol. 6383, pp. 578–583. Springer, Berlin/Heidelberg/New York (2010). EC-TEL

Argote, L.: Reflections on two views of managing learning and knowledge in organizations. J. Manag. Inq. **14**(1), 43–48 (2005)

Argote, L., McElvily, B., Reagans, R.: Managing knowledge in organizations: an integrative framework and review of emerging themes. Manag. Sci. **49**(4), 571–582 (2003)

Argyris, C.: On Organizational Learning. Oxford University Press, Oxford/New York (1992)

Argyris, C.: Reasons and Rationalizations: The Limits to Organizational Knowledge. Oxford University Press, Oxford/New York (2004)

Armistead, C., Meakins, M.: A framework for practising knowledge management. Long Range Plann. **35**(1), 49–71 (2002)

Axsater, S.: A note on stock replenishment and shipment scheduling for vendor-managed inventory systems. Manag. Sci. **47**(9), 1306–1310 (2001)

Baldridge, D. C., Floyd, S. W. and Markóczy, L.: Are managers from Mars and academicians from venus? Toward an understanding of the relationship between academic quality and practical relevance. Strat. Mgmt. J. **25**, 1063–1074. doi: 10.1002/smj.406 (1997)

Barney, J.: Firm resources and sustained competitive advantage. J. Manag. **17**(1), 99–120 (1991)

Bate, S.: Strategies for Cultural Change. Oxford Publications/Butterworth Heinemann, New York (1994)

Behling, O., Eckel, N.: Making sense out of intuition. Acad. Manag. Exec. **5**(1), 46–54 (1991)

Belcher, L.: Leveraging supply chain execution strategies. Stores. **81**(1), (2008)

Bititci, U.S.: Dynamics of performance measurement systems. Int. J. Oper. Manag. **20**(6), 692–704 (2000)

K. Ganesh et al., *Design and Development of Knowledge Management for Manufacturing: Framework, Solution and Strategy*, DOI 10.1007/978-3-319-02892-7,

Bixler, C.: Codification, personalization, and integration. KMWorld **11**(3), 18–20 (2002)

Blatherwick, A.: Vendor-managed inventory: fashion fad or important supply chain strategy? Supply Chain Manag. Int. J. **3**(1), 10–11 (1998)

Boisot, M.: Knowledge assets: securing competitive advantage in the information economy. Oxford University Press, New York (1999)

Bontis, N., Crossan, M.M., Hulland, J.: Managing an organizational learning system by aligning stocks and flows. J. Manag. Stud. **39**(4), 437–469 (2002)

Brandt, D.: Supply chain dream come true. Ind. Eng. **41**(9), 52–53 (2009)

Brown, S., Squire, B., BlacKMon, K.: The contribution of manufacturing strategy involvement and alignment to world-class manufacturing performance. Int. J. Oper. Prod. Manag. **27**(3), 282–302 (2007)

Bui, T.X., Sankaran, S., Sebastian, I.M.: A framework for measuring national e-readiness. Int. J. Electron. Bus. **1**(1), 3–22 (2003)

Carbone, J.: Use your purchasing clout to get better customer service. Purchasing **4**(1), 57–59 (2009)

Carthy, I.M., Ridgway, K.: Cladistics: a taxonomy for manufacturing organizations. Integr. Manuf. Syst. **11**(1), 16–29 (2000)

Cavaleri, S.A.: Leveraging organizational learning for knowledge performance. Learn. Organ. **11**(2), 159–176 (2004)

Cetinkaya, S., Lee, C.Y.: Stock replenishment and shipment scheduling for vendor-managed inventory systems. Manag. Sci. **46**(2), 217–231 (2000)

Chaouch, B.A.: Stock levels and delivery rates in vendor managed inventory programs. Prod. Oper. Manag. **10**(1), 31–44 (2001)

Chavan, M.: The balanced scorecard: a new challenge. J. Manag. Dev. **28**(5), 393–406 (2009)

Chen, M.F.: The joint impact of relationship-selling behaviors and switching barriers in professional service contexts. Int. J. Serv. Oper. Inform. **4**(2), 123–145 (2009)

Chen, C.J., Huang, J.W.: How organizational climate and structure affect knowledge management – the social interaction perspective. Int. J. Inf. Manag. **27**(2), 104–118 (2007)

Chen, L., Mohamed, S.: Contribution of knowledge management activities to organizational business performance. J. Eng. Des. Technol. **6**(3), 269–285 (2008)

Chen, T.Y., Chen, C.B., Peng, S.Y.: Firm operation performance analysis using data envelopment analysis and balanced scorecard: a case study of a credit cooperative bank. Int. J. Product. Perform. Manag. **57**(7), 523–539 (2008)

Chen, S.H., Wang, H.H., Yang, K.J.: Establishment and application of performance measure indicators for universities. TQM J. **21**(3), 220–235 (2009)

Chia, A., Goh, M., Hum, S.H.: Performance measurement in supply chain entities: balanced scorecard perspective. Benchmark. Int. J. **16**(5), 605–620 (2009)

Choi, B., Lee, H.: Knowledge management strategy and its link to knowledge creation process. Expert Syst. Appl. **23**(3), 173–187 (2002)

Choi, I., Jung, J., Song, M.: A framework for the integration of knowledge management and business process management. Int. J. Innov. Learn. **1**(4), 399–408 (2004)

Choo, A.S., Linderman, K., Schroeder, R.G.: Method and context perspectives on learning and knowledge creation in quality management. J. Oper. Manag. **25**(4), 918–931 (2007a)

Choo, A.S., Linderman, K.W., Schroeder, R.G.: Method and psychological effects on learning behaviors and knowledge creation in quality improvement project. Manag. Sci. **53**(3), 437–451 (2007b)

Chopra, S., Meindl, P.: Supply Chain Management: Strategy, Planning and Operations, 3rd edn. Pearson International Edition, London (2007)

Chua, A.: Knowledge management system architecture a bridge between KM consultants and technology. Int. J. Inf. Manag. **24**(1), 87–98 (2004)

Claassen, M., Weele, A., Raaij, E.: Performance outcomes and success factors of vendor managed inventory (VMI). Supply Chain Manag. **13**(6), 406–414 (2008)

Cohen, S., Thiraios, D., Kandilorou, M.: Performance parameters interrelations from a balanced scorecard perspective: an analysis of Greek companies. Manag. Audit. J. **23**(5), 485–503 (2008)

Cronbach, L.J.: Coefficient alpha and the internal structure of tests. Psychometrika **16**(3), 297–334 (1951)
Croson, R., Donohue, K.: Upstream versus downstream information and its impact on the bullwhip effect. Syst. Dyn. Rev. **21**(3), 249–260 (2005)
Crossan, M.M., Lane, H.W., White, R.E.: An organizational learning framework: from intuition to institution. Acad. Manag. Rev. **24**(3), 522–537 (1999)
Daugherty, P.J., Myers, M.B., Wautry, C.: Automatic replenishment programs: an empirical examination. J. Bus. Logist. **20**(2), 63–82 (2001)
David, I., Eben-Chaime, M.: How far should JIT vendor-buyer relationships go? Int. J. Prod. Econ. **8**(2), 361–368 (2003)
De Kereki, I.F., Azpiazu, J., Silva, A.: Knowledge management in learning environment design. Front. Educ. **3**, S3F6–S3F11 (2004)
De Toni, A.F., Zamolo, E.: From a traditional replenishment system to vendor-managed inventory: a case study from the household electrical appliances sector. Int. J. Prod. Econ. **96**(1), 16–19 (2004)
DeMeuse, K.P., McDaris, K.K.: An exercise in managing change. Train. Dev. J. **48**(2), 55–57 (1994)
Disney, S.M., Towill, D.R.: A procedure for the optimization of the dynamic response of a vendor managed inventory system. Comput. Ind. Eng. **43**(2), 27–58 (2002)
Disney, S.M., Towill, D.R.: Vendor managed inventory and bullwhip reduction in a two level supply chain. Int. J. Oper. Prod. Manag. **23**(6), 625–651 (2003a)
Disney, S.M., Towill, D.R.: The effect of vendor managed inventory (VMI) dynamics on the bullwhip effect in supply chains. Int. J. Prod. Econ. **85**(2), 199–215 (2003b)
Disney, S.M., Naim, M.M., Potter, A.T.: Assessing the impact of e-business on supply chain dynamics. Int. J. Prod. Econ. **89**(2), 109–118 (2004)
Doherty, M., Lau, M.S., Kaur, S., Jain, R.: Using ontology to improve knowledge management in technology transfer. Int. J. Technol. Transf. Commer. **4**(1), 105–125 (2005)
Dong, Y., Xu, K.: A supply chain model of vendor-managed inventory. Transp. Res. Part E Logist. Transp. Rev. **38**(2), 75–95 (2002)
Dresner, M., Yao, Y., Dong, Y.: Managing supply chain backorders under vendor managed inventory: an incentive approach and empirical analysis. Eur. J. Oper. Res. **203**(2), 350–362 (2009)
Dyck, B., Starke, F.A., Miscke, G.A., Mauws, M.: Learning to build a car: an empirical investigation of organizational learning. J. Manag. Stud. **42**(2), 387–416 (2005)
Edward, M.M.: Building trust at the speed of change: the power of the relationship based corporations. American Management Association (AMACOM), New York (2000)
Franken, A., Braganza, A.: Organizational forms and knowledge management: one size fits all. Int. J. Knowl. Manag. Stud. **1**(½), 18–37 (2006)
Friedman, V.J., Lipshitz, R., Popper, M.: The mystification of organizational learning. J. Manag. Inq. **14**(1), 19–30 (2005)
Fry, M.J., Kapuscinski, R., Olsen, T.L.: Coordinating production and delivery under a (z, Z)-type vendor managed inventory contract. Manuf. Serv. Oper. Manag. **3**(2), 151–173 (2001)
Gamble, P.R., Blackwell, J.: Knowledge Management. Kogan Page, London (2001)
Gao, Y., Wei, W.: Are corporate political actions ethical? An investigation of executives in Chinese enterprises. Int. J. Bus. Innov. Res. **1**(4), 464–480 (2007)
Gentry, C.: Retailers and suppliers get linked. Chain Store Age **84**, 54 (2008)
Gerchak, Y., Khmelnitsky, E.: A consignment system where suppliers cannot verify retailer's sales reports. Int. J. Prod. Econ. **83**(1), 37–43 (2003)
Germain, R., Iyer, K.N.: The interaction of internal and downstream integration and its association with performance. J. Bus. Logist. **27**(2), 29–52 (2006)
Goncalves, H.S.: Proposal of a strategy model planning aligned to the balanced scorecard and the quality environments. TQM J. **21**(5), 462–472 (2009)
Green, A., Ryan, J.C.H.: A framework of intangible valuation areas (FIVA): aligning business strategy and intangible assets. J. Intellect. Cap. **6**(1), 43–52 (2005)
Gunther, C.W., Rinderle-ma, S., Reichert, M., Van Der Aalst, W.M.P., Recker, J.: Using process mining to learn from process change in evolutionary. Int. J. Bus. Process. Integr. Manag. **3**(1), 61–78 (2008)

Gupta, B., Sharma, N.K., Ganesh, C.: The influence of organizational cultural values, reward, time, self-esteem and job security on knowledge sharing intentions among managers. Int. J. Indian Cult. Bus. Manag. **2**(2), 125–143 (2009)

Guthrie, J., Yongvanich, K.: Balanced scorecard practices amongst Thai companies: performance effects. Pac. Account. Rev. **21**(2), 132–149 (2009)

Haeckel, S.H., Nolan, R.L.: Managing by wire. Harv. Bus. Rev. **71**(5), 122–132 (1993)

Hair, J.F., Anderson, R.E., Tatham, R.L., Black, W.C.: Multivariate Data Analysis, 5th edn. Macmillan, New York (1998)

Hanafizadeh, P., Rezaei, M., Ghafouri, A.: Defining strategic processes in investment companies: an exploration study in Iranian investment companies. Bus. Process. Manag. J. **15**(1), 20–33 (2009)

Hans, J.P.: Learning from the past and knowledge management: are we making progress? J. Loss Prev. Process. Ind. **22**(6), 672–679 (2009). Elsevier

Hansen, M., Nohria, N., Tierney, T.: What's your strategy for managing knowledge? Harv. Bus. Rev. **77**(2), 106–116 (1999)

Hassan, A., Dantan, J.Y., Siadat, A.: Information modelling for variation risk management during product and process design. Int. J. Prod. Qual. Manag. **2**(4), 221–240 (2007)

Hayes, R.H., Wheelwright, S.C.: Restoring Our Competitive Edge: Competing Through Manufacturing. Wiley, New York (1984)

Hazlett, S.A., McAdam, R., Gallagher, S.: Theory building in knowledge management: in search of paradigms. J. Manag. Inq. **14**(1), 31–42 (2005)

Holmstrom, J.: Business process innovation in the supply chain – a case study of implementing vendor managed inventory. Eur. J. Purch. Suppl. Manag. **4**(3), 127–131 (1998)

Holsapple, C.W., Joshi, K.D.: Knowledge manipulation activities: results of a Delphi study. J. Inf. Manag. **39**(6), 477–490 (2002)

Holt, D.T., Bartczak, S.E., Clark, S., Trent, M.R.: The development of an instrument to measure readiness for knowledge management. Knowl. Manag. Res. Pract. **5**(1), 75–92 (2007)

Hoque, M.A., Goyal, S.K.: An optimal policy for a single-vendor single-buyer integrated production-inventory system with capacity constraint of the transport equipment. Int. J. Prod. Econ. **65**(3), 305–315 (2000)

Huang, H. L.: User acceptance of mobile knowledge management learning system: design and analysis. Educ. Technol. Soc. **13**(3), 70–77 (2010)

Huang, B., Ye, Z.: The effects of lumpy demand and shipment size constraint: a response to revisit the note on supply chain integration in vendor-managed inventory. Decis. Support. Syst. **48**(2), 421–425 (2010)

Hurley, T.A., Green, C.W.: Creating a knowledge management culture: the role of task, structure, technology, and people in encouraging knowledge creation and transfer. In: Proceedings. (2005)

Hwang, H.G., Chiu Chang, I., Chen, F.J., Ying Wu, S.: Investigation of the application of KMS for diseases classifications: a study in a Taiwanese hospital. Expert Syst. Appl. **34**(1), 725–733 (2008)

Isaacs, W.N.: Dialogue, collective thinking and organizational learning. Organ. Dyn. **22**(2), 24–39 (1993)

Islam, N.: Innovative manufacturing readiness levels (IMRLs): a new readiness matrix. Int. J. Nanomanuf. **6**(1–4), 362–375 (2010)

James, W.: Rationality, institutionalism and accounting change: understanding a performance management system within an Australian public sector entity. J. Account. Organ. Chang. **5**(3), 362–389 (2009)

James, A.S., Abraham, B.R.S.: Integrating knowledge management, learning mechanisms, and company performance. In: Practical Aspects of Knowledge Management, vol. 2586, pp. 620–631. Springer, New York (2002). no. 25

Jaruphongsa, W., Cetinkaya, S., Lee, C.Y.: Warehouse space capacity and delivery time window considerations in dynamic lot sizing for a simple supply chain. Int. J. Prod. Econ. **92**(2), 169–180 (2003)

Kaipia, R., Holmstrom, J., Tanskanen, K.: VMI: what are you losing if you let your customer place orders? Prod. Plan. Control **13**(1), 17–25 (2002)

Kalling, T.: ERP systems and the strategic management processes that lead to competitive advantage. Inf. Resour. Manag. J. **16**(4), 46–67 (2003)

Kauremaa, J., Smaros, J., Holmstrom, J.: Patterns of vendor-managed inventory: findings from a multiple-case study. Int. J. Oper. Prod. Manag. **29**(11), 1109–1139 (2009)

Keith, M., Goul, M., Demirkan, H., Nichols, J., Mitchell, M.C.: Contextualising knowledge management readiness to support change management strategies. In: Proceedings of the 39th Annual Hawaii International Conference on System Sciences (HICSS'06) Track 7. (2006)

Kenett, R.S., Raphaeli, O.: Multivariate methods in enterprises system implementation, risk management and change management. Int. J. Risk Assess. Manag. **9**(3), 258–276 (2008)

Kenneth, D.S.: Comparing learning and knowledge management theories in an Australian telecommunications practice. Asian J. Manag. Case. **7**(1), 33–54 (2010)

Khalfan, M.M.A., Anumba, C.J., Siemieniuch, C.E., Sinclair, M.A.: Readiness assessment of the construction supply chain for concurrent engineering. Eur. J. Purch. Suppl. Manag. **7**(2), 141–153 (2001)

Khalfan, M.A., Kashyap, M., Li, X., Abbot, C.: Knowledge management in construction supply chain integration. Int. J. Netw. Virt. Organ. **7**(2/3), 207–221 (2010)

Kiesmüller, G.P., BroeKMeulen, R.A.C.M.: The benefit of VMI strategies in a stochastic multi-product serial two echelon system. Comput. Oper. Res. **37**(2), 406–416 (2010)

King, W.R., Marks Jr., P.V.: Motivating knowledge sharing through a knowledge management system. Omega **36**(1), 131–146 (2008)

Kittiya, Y., Guthrie, J.: Balanced scorecard practices amongst Thai companies: performance effects. Pac. Account. Rev. **21**(2), 132–149 (2009)

Kleywegt, A.J., Nori, V.S., Savelsbergh, M.W.P.: The stochastic inventory routing problem with direct deliveries. Transp. Sci. **36**(1), 94–118 (2002)

KPMG: Knowledge Management Research Report. (1998)

Krishnan, R.: Knowledge management and competitive advantage. IIMB Manag. Rev. **13**(4), 23–38 (2001)

Kuk, G.: Effectiveness of vendor-managed inventory in the electronics industry: determinants and outcomes. Inf. Manag. **41**(5), 645–654 (2004)

Kulp, S.C.: The effect of information precision and information reliability on manufacturer-retailer relationships. Account. Rev. **77**(3), 653–677 (2002)

Kunz, J., Linder, S.: The impact of management control systems on knowledge process- a behavioral perspective. Int. J. Strateg. Chang. Manag. **2**(1), 54–72 (2010)

Lai, H.M.S., Sculli, D.: A taxonomy for industrial network strategies. Int. J. Intercult. Inf. Manag. **1**(3), 275–299 (2009)

Lansiluoto, A., Jarvenpaa, M.: Environmental and performance management forces: integrating 'greenness' into balanced scorecard. Qual. Res. Acc. Manag. **5**(3), 184–206 (2008)

Larsson, J., Backstrom, I., Wiklund, H.: Leadership and organizational behavior – similarities between three award winning organizations. Int. J. Manag. Pract. **3**(4), 327–345 (2009)

Leonard-Barton, D.: Wellsprings of Knowledge. Harvard Business School Press, Boston (1995)

Leseure, M.: Organizational design alternatives within international operations networks: a transaction cost perspective. Int. J. Serv. Oper. Manag. **7**(4), 419–439 (2010)

Leung, S., Wong, W., Qi, J.: Coordinating supply chains with sales rebate contracts and vendor-managed inventory. Int. J. Prod. Econ. **120**(1), 151–159 (2009)

Levitt, B., March, J.G.: Organizational learning. Annu. Rev. Sociol. **14**(1), 319–340 (1988)

Lewin, L.: Field theory and experiments in social psychology. Am. J. Sociol. **44**, 879 (1939)

Lewin, K., Lippitt, R., White, R.: Patterns of aggressive behaviors in experimentally created social climates. J. Soc. Psychol. **10**(2), 271–293 (1939)

Li, J.H., Raym, P., Chattopadhyaya, S.: E-Health readiness framework from electronic health records perspective. Int. J Int. Enterp. Manag. **6**(4), 326–348 (2010)

Lieb, R., Bentz, B.A.: The use of third-party logistics services by large American manufacturers: the 2004 survey. Transp. J. **44**(2), 5–15 (2005)

Lissitz, R.W., Green, S.B.: Effect of the number of scale points on reliability: a Monte Carlo approach. J. Appl. Psychol. **60**(1), 10–13 (1975)
Liu, Y., Loh, H.T., Toumi, K.Y., Tor, S.B.: A hierarchical text classification system for manufacturing knowledge management and retrieval. Int. J. Knowl. Manag. Stud. **2**(4), 406–425 (2008)
Lodree Jr., E.J., Geiger, C.D., Jiang, X.: Taxonomy for integrating scheduling theory and human factors: review and research opportunities. Int. J. Ind. Ergon. **39**(1), 39–51 (2009)
Loss, L., Klen, A.A.P., Rabelo, R.J.: A framework for learning collaborative networked organizations. Int. J. Serv. Oper. Manag. **6**(3), 313–334 (2010)
Lovrich, N.P., Pierce, J.C.: Knowledge gap phenomena: effect of situation-specific and transsituational factors. Commun. Res. **11**(3), 415–434 (1984)
Lu, L.: A one-vendor multi-buyer integrated inventory model. Eur. J. Oper. Res. **81**(2), 312–323 (1995)
Lynch, A., Sheahan, C.: Taxonomy of a decision: ranking operational decisions within the SME. Int. J. Decis. Sci. **1**(1/2), 126–141 (2009)
MAKE: Survey of Most Admired Knowledge Enterprises, (MAKE). (1998)
Marloes, J.T.C., Arjan, J., Van, W., Erik, M., Van, R.: Performance outcomes and success factors of vendor managed inventory. Suppl. Chain Manag. Int. J. **13**(6), 406–414 (2008)
Marquardt, M.J.: Action Learning in Action. Davies Black, Palo Alto (1999)
Masataka, T., Yusuke, H., Mitsuru, I., Riichiro, M.: Environment to integrate practice and learning based on collaborative space ontology and patterns. In: Intelligent Tutoring Systems, vol. 4053, no. 1, pp. 187–196. Springer (2006)
Michaelraj, L., Shahabudeen, P.: Replenishment policies for sustainable business development in a continuous credit based vendor managed inventory distribution system. Comput. Ind. Eng. **56**(1), 260–266 (2009)
Mills, A.M., Smith, T.A.: Knowledge management and organizational performance: a decomposed view. J. Knowl. Manag. **15**(1), 156–171 (2011)
Miri-Nargesi, S.S., Keramati, A., Haleh, H., Ansarinejad, A.: Assessing organizational readiness to implement customer relationship management using fuzzy analytical network process: a case study. Int. J. Acad. Res. **3**(3), 6–9 (2011)
Mishra, B., Raghunathan, S.: Retailer vs vendor-managed inventory and brand competition. Manag. Sci. **50**(4), 445–457 (2004)
Mohammadi, K., Khanlari, A., Sohrabi, B.: Organizational readiness assessment for knowledge management. Int. J. Knowl. Manag. **5**(1), 29–45 (2009)
Moon, G., Park, C., Yoo, D.K., Choi, S.: Effective implementation of communities of practices (Cops) in a knowledge habitat: a case study of Samsung electronics. Int. J Serv. Oper. Manag. **8**(3), 335–346 (2011)
Moutinho, L., Hutcheson, G.D.: Store choice and patronage: a predictive modelling approach. Int. J. Bus. Innov. Res. **1**(3), 233–252 (2007)
Mrayyan, M.T., Modallal, R., Awamreh, K., Atoum, M., Abdullah, M., Suliman, S.: Readiness of organizations for change, motivation and conflict-handling intentions: senior nursing student's perceptions. Nurs. Educ. Pract. **8**(2), 120–128 (2008)
Nadler, D., Tushman, M.: Strategic Organization Design: Concepts, Tools and Processes. Scott Foresman, Glenview/London (1988)
Nelson, D., Quick, J.: Organizational Behavior, pp. 558–560. West Publishing, New York (1994)
Nevo, D., Chan, Y.E.: A Delphi study of knowledge management systems: scope and requirements. Inf. Manag. **44**(6), 583–597 (2007a)
Nevo, D., Chan, Y.E.: A temporal approach to expectations and desires from knowledge managements systems. Decis. Support. Syst. **44**(1), 298–312 (2007b)
Niehoff, B.P., Moorman, R.H.: Exploring the relationships between top management behaviors and employee perceptions of fairness. Int. J. Publ. Adm. **19**(6), 941–961 (1996)
Nielsen, A.P.: Understanding dynamic capabilities through knowledge management. J. Knowl. Manag. **10**(4), 59–71 (2006)
Nonaka, I.: The knowledge-creating company. Harv. Bus. Rev. **69**(6), 96–104 (1991). November-December

Nonaka, I.: A dynamic theory of organizational knowledge creation. Organ. Sci. **5**(1), 14–37 (1994)

Nonaka, I., Takeuchi, H.: The Knowledge Creating Company. Oxford University Press, New York (1995)

Nonaka, I., Toyama, R., Konno, N.: SECI, Ba and leadership: a unified model of dynamic knowledge creation. Long Range Plan. **33**(1), 5–34 (2002)

Nunnally, J.C.: Psychometric Theory, 2nd edn. McGraw-Hill, New York (1978)

Organ, D.W., Moorman, R.H.: Fairness and organizational citizenship behavior: what are the connections? Soc. Justice Res **6**(1), 5–18 (1993)

Örtenblad, A.: On differences between organizational learning and learning organization. Learn. Org. **8**(3), 125–133 (2001)

Pablos, P.O.: Human resource management systems and knowledge management: the key to develop interorganizational teams. Int. J. Learn. Intellect. Cap. **1**(2), 121–131 (2004)

Pagell, M.: Understanding the factors that enable and inhibit the integration of operations, purchasing and logistics. J. Oper. Manag. **22**(5), 459–487 (2004)

Paiva, L., Roth Aleda, V., Fensterseifer, J.E.: Organizational knowledge and the manufacturing strategy process: a resource-based view analysis. J. Oper. Manag. **26**(1), 115–132 (2008)

Pedler, M., Boydell, T., Burgoyne, J.: Towards the learning company. Manag. Educ. Dev. **20**(1), 1–8 (1989)

Persaud, A.: The knowledge gap. Foreign Aff. **80**(2), 107–117 (2001)

Peter, Y.T.S., John, L.S.: Process Level Integration of Organizational Learning, Learning Organization and Knowledge Management, vol. 2, no. 3/4, pp. 308–319. Wiley International, UK (2006)

Piplani, R., Viswanathan, S.: A model for evaluating supplier-owned inventory strategy. Int. J. Prod. Econ. **81**(2), 565–571 (2003)

Podsakoff, P.M., MacKenzie, S.B., Moorman, R.H., Fetter, R.: Transformational leader behaviors and their effects on followers trust in leader, satisfaction, and organizational citizenship behaviors. Leadersh. Q. **1**(2), 107–142 (1990)

Prahalad, C.K., Hamel, G.: The core competence of the corporation. Harv. Bus. Rev. **68**, 1–15 (1990)

Prietula, M.J., Simon, H.A.: The experts in your midst. Harv. Bus. Rev. **67**(1), 120–124 (1989)

Punniyamoorthy, M., Murali, R.: Balanced score for the balanced scorecard: a benchmarking tool. Benchmark. Int. J. **15**(4), 420–443 (2008)

Quinn, J.B.: The intelligent enterprise: a new paradigm. Acad. Manag. Exec. **6**(4), 48–63 (1992)

Rahim, M.A.: Toward a theory of managing organizational conflict. Int. J. Confl. Manag. **13**(3), 206–235 (2002)

Raman and Murali: Designing knowledge management systems for teaching and learning with wiki technology. J. Inf. Syst. Edu. **53**(53), 171–283 (2000)

Ramrakhyani, R.: Analysis of vendor managed consignment inventory systems. State University of New York at Binghamton, Proquest (2009)

Rautiainen, A.I.: The interrelations of decision-making rationales around BSC adoptions in Finnish municipalities. Int. J. Product. Perform. Manag. **58**(8), 787–802 (2009)

Reddi, K.R., Moon, Y.B.: A framework for engineering change management in enterprise resource planning using service oriented architecture. Int. J. Bus. Inf. Syst. **8**(1), 46–65 (2011)

Reddy, M., Vrat, P.: Vendor managed inventory model: a case study. J. Adv. Manag. Res. **4**(1), 83–88 (2007)

Rehman, F.U., Yan, X.T.: Supporting early design decision making using design context. Int. J. Des. Res. **6**(1/2), 169–189 (2007)

Revilla, E., Sarkis, J., Acosta, J.: Towards a knowledge management and learning taxonomy for research joint ventures. Technovation **25**(11), 1307–1316 (2005)

Robinson, V.M.J.: Descriptive and normative research on organizational learning: locating the contribution by Argyris and Schon. Int. J. Edu. Manag. **15**(2), 58–67 (2001)

Roth, A.V., Miller, J.G.: Manufacturing Strategy, Manufacturing Strength, Managerial Success, and Economic Outcomes, pp. 97–108. Kluwer Academic, Norwell (1990)

Ruben, B.D., Kealey, D.: Behavioral assessment of communication competency and the prediction of cross-cultural adaptation. Int. J. Int. Relat. **3**(1), 15–47 (1979)

Ryoo, J., Oh, T.H., Shin, S., Choi, Y.B.: A comprehensive readiness assessment framework for identify theft safeguards in web-based electronic government systems. Int. J. Electron. Gov. **6**(1), 19–40 (2009)

Sabherwal, R., Fernandez, I.B.: An empirical study of the effect of knowledge management processes at individual, group, and organizational levels. Decis. Sci. **34**(2), 225–261 (2003)

Salah, F., Abdulla, H.: A knowledge based system for enhancing conceptual design. Int. J. Comput. Appl. Technol. **40**(1/2), 23–36 (2011)

Sari, K.: Exploring the benefits of vendor managed inventory. Int. J. Phys. Distrib. Logist. Manag. **37**(7), 529–545 (2007)

Saxena, R.: Vendor-managed inventory. Industrial Engineer, Proquest, (2009)

Seely-Brown, J., Duguid, P.: Organizational learning and communities-of-practice: toward a unified view of working, learning and innovation. Organ. Sci. **2**(1), 40–57 (1991)

Senge, P.: The Fifth Discipline: The Art and Practice of the Learning Organization. Doubleday/Currency, New York (1990)

Shirazi, A., Mortazavi, S., Azad, N.P.: Factors affecting employees readiness for knowledge management. Eur. J. Econ. Financ. Adm. Sci. **7**(33), 167–177 (2011)

Siemieniuch, C.E., Sinclair, M.A.A.: A framework for organizational readiness for knowledge management. Int. J. Oper. Prod. Manag. **24**(1), 79–98 (2004)

Singh, S.K.: Knowledge management practices and organizational learning in Indian software company. Int. J. Bus. Innov. Res. **3**(4), 363–381 (2009)

Smaros, J., Holmstrom, J.: Viewpoint: reaching the consumer through e-grocery VMI. Int. J. Retail. Distrib. Manag. **28**(3), 55–60 (2000)

Smaros, J., Lehtonen, J.M., Appelqvist, P., Holmstrom, J.: The impact of increasing demand visibility on production and inventory control efficiency. Int. J. Phys. Distrib. Logist. Manag. **33**(4), 336–354 (2003)

Snyder, C.A., McManus, D.J., Wilson, L.T.: Corporate memory management: a knowledge management process model. Int. J. Technol. Manag. **20**(5–8), 752–764 (2000)

Sousa, C.A.A., Hendriks, P.H.J.: The driving bell and the butterfly: the need for grounded theory in developing a knowledge-based view of organization. Organ. Res. Method. **9**(3), 315–338 (2006)

Spiegler, I.: Knowledge management: a new idea or a recycled concept? Commun. Assoc. Inf. Syst. **14**(3), 1–14 (2000)

Spiteri, C., Borg, J.C.: System architecture for mobile knowledge management within product lifecycle design. Int. J. Manuf. Res. **5**(4), 396–412 (2010)

Sun, P.Y.T., Scott, J.L.: Exploring the divide: organizational learning and learning organization. Learn. Organ. **10**(4), 202–215 (2003)

Suppiah, V., Sandhu, M.S.: Organizational culture's influence on tacit knowledge sharing behavior. J. Knowl. Manag. **15**(3), 462–477 (2011)

Tanskanen, K., Holmstrom, J., Elfving, J., Talvitie, U.: Vendor-managed-inventory (VMI) in construction. Int. J. Product. Perform. Manag. **58**(1), 29–40 (2008)

Thakkar, J., Kanda, A., Deshmukh, S.G.: Supply chain performance measurement framework for small and medium scale enterprises. Benchmark. Int. J. **16**(5), 702–723 (2009)

Tsang, E.W.K.: Organizational learning and the learning organization: a dichotomy between descriptive and prescriptive research. Hum. Relat. **50**(1), 73–89 (1997)

Tyan, J., Wee, H.M.: Vendor managed inventory: a survey of the Taiwanese grocery industry. J. Purch. Supply Manag. **9**(1), 11–18 (2003)

Varma, S., Wadhwa, S., Deshmukh, S.G.: Evaluating petroleum supply chain performance: application of analytical hierarchy process to balanced scorecard. Asia Pac. J. Mark. Logist. **20**(3), 343–356 (2008)

Vera, D., Crossan, M.M.: Strategic leadership and organizational learning. Acad. Manag. Rev. **29**(2), 222–240 (2004)

Vigtil, A.: Information exchange in vendor managed inventory. Int. J. Phys. Distrib. Logist. Manag. **37**(2), 131–147 (2007)

Viswanathan, S., Piplani, R.: Coordinating supply chain inventories through common replenishment epochs. Eur. J. Oper. Res. **129**(2), 277–286 (2001)

Viswanathan, S., Wang, Q.: Discount pricing decisions in distribution channels with price-sensitive demand. Eur. J. Oper. Res. **149**(3), 571–587 (2003)

VMI. Available at http://www.vendormanagedinventory.com

Von Krogh, G., Nonaka, I., Aben, M.: Making the most of your company's knowledge: a strategic framework. Long Range Plan. **34**(4), 421–439 (2001)

Vouzas, F.K.: Business excellence and human resources: investigating best practices in European quality awarded Hellenic organizations. Int. J. Bus. Innov. Res. **3**(3), 281–297 (2009)

Waller, M., Johnson, M.E., Davis, T.: Vendor-managed inventory in the retail supply chain. J. Bus. Logist. **20**(1), 183–203 (1999)

Wang, Y., Ru, J.: Consignment contracting: who should control inventory in the supply chain? Eur. J. Oper. Res. **201**(3), 760–766 (2009)

Wei, X., Xie, F.: Knowledge management processes and innovation: an empirical analysis of firms in software cluster. Int. J. Hum. Resour. Dev. Manag. **8**(½), 25–42 (2008)

Wild, R.H., Griggs, K.A., Downing, T.: A framework for e-learning as a tool for knowledge management. Ind. Manag. Data Syst. **102**(7), 371–380 (2002)

Xu, K., Leung, M.: Stocking policy in a two-party vendor managed channel with space restrictions. Int. J. Prod. Econ. **117**(2), 271–277 (2009)

Yang, J.: Integrative performance evaluation for supply chain system based on logarithm triangular fuzzy number-AHP method. Kybernetes **1**(10), 1760–1770 (2009)

Yang, K.K., Ruben, R.A., Webster, S.: Managing vendor inventory in a dual level distribution system. J. Bus. Logist. **24**(2), 91–105 (2003)

Zack, M.H.: Developing a knowledge strategy. Calif. Manag. Rev. **41**(3), 125–145 (1999). Springer

Zammori, F., Marcello, B., Frosolini, M.: A standard agreement for vendor managed inventory. Strateg. Outsourcing Int. J. **2**(2), 165–186 (2009)

Zhao, Q.H., Cheng, T.: An analytical study of the modification ability of distribution centers. Eur. J. Oper. Res. **194**(3), 901–910 (2009)

Zhao, B., Tian, Z.: Relationship between consumer ethics and social rewards-punishments in mainland China. Int. J. Bus. Innov. Res. **3**(1), 52–69 (2009)

Zhao, X., Sum, C., Qi, Y., Zhang, H., Lee, T.: A taxonomy of manufacturing strategies in China. J. Oper. Manag. **24**(5), 621–636 (2006)

Index

K. Ganesh et al., *Design and Development of Knowledge Management for Manufacturing: Framework, Solution and Strategy*, DOI 10.1007/978-3-319-02892-7,

MIX
Papier aus verantwortungsvollen Quellen
Paper from responsible sources
FSC® C105338

If you have any concerns about our products,
you can contact us on
ProductSafety@springernature.com

In case Publisher is established outside the EU,
the EU authorized representative is:
Springer Nature Customer Service Center GmbH
Europaplatz 3, 69115 Heidelberg, Germany

Printed by Libri Plureos GmbH
in Hamburg, Germany